S294 Cell biology

Book 3
Challenging Cells

This publication forms part of the Open University module S294 *Cell biology*. The complete list of texts which make up this module can be found at the back. Details of this and other Open University modules can be obtained from the Student Registration and Enquiry Service, The Open University, PO Box 197, Milton Keynes MK7 6BJ, United Kingdom (tel. +44 (0)845 300 60 90; email general-enquiries@open.ac.uk).

Alternatively, you may visit the Open University website at www.open.ac.uk where you can learn more about the wide range of modules and packs offered at all levels by The Open University.

To purchase a selection of Open University materials visit www.ouw.co.uk, or contact Open University Worldwide, Walton Hall, Milton Keynes MK7 6AA, United Kingdom for a catalogue (tel. +44 (0)1908 274066; fax +44 (0)1908 858787; email ouw-customer-services@open.ac.uk).

The Open University, Walton Hall, Milton Keynes MK7 6AA

First published 2012. Second edition 2014

Edited and designed by The Open University.

Typeset by The Open University.

Printed and bound in the United Kingdom by Halstan Printing Group, Amersham.

ISBN 978 1 7800 7879 3

2.1

Contents

Chapter 1   The life cycle of the cell

1.1   Introduction

The final book of the module looks at some of the ways in which cells respond to internal and external challenges. What effect do nutrients and environmental conditions have on the ability of cells to survive and multiply? How does damage in the form of infection or genetic defects result in disease, and how can the properties of cells be manipulated to improve human health? The topics in this book bring together many of the concepts, and much of the knowledge, that you have acquired in your studies of cell biology so far. This first chapter explores how cells undergo proliferation, differentiation or death in respond to external and internal cues, and how these processes contribute to the development of complex multicellular organisms.

All cells are generated by the division of other cells. Unicellular organisms, such as bacteria, divide when conditions are appropriate to form two new individuals. In contrast, the growth and development of a multicellular organism requires a highly regulated programme of cell division and differentiation to generate diverse cell types. Many of the tissues of a mature multicellular organism must be continuously renewed throughout life in order to maintain the organism's form and function. Regulated cell death also plays a crucial role, not only in ensuring that old and damaged cells are removed, but also in maintaining appropriate cell numbers and determining the morphology of the developing organism.

How is cell division controlled so that the appropriate numbers of cells are produced? What generates cell diversity? Why and how do some cells die, and what enables other cells to live for many years, only dying when the organism dies?

There is not enough space here to answer these complex questions in detail, so this chapter will present an overview, using examples drawn mostly from the animal kingdom, to illustrate some of the general principles governing the regulation of cell proliferation, differentiation and death. Your study of these regulatory processes will bring together a number of concepts from Books 1 and 2 of the module, including the roles of signalling pathways, protein activation, gene expression and internal cell movement. Although cell proliferation, differentiation and death will be considered in separate sections of the chapter, it should become clear to you that these processes are in fact inextricably linked.

1.2   The regulation of cell proliferation

Cells increase in number as a result of cell growth followed by division. The control of these processes is one of the most widely studied areas of cell biology. As well as the fundamental role of cell growth and division in the reproduction and development of organisms, abnormal cell proliferation is the cause of a number of diseases, including cancer.

You will recall from Book 1, Section 4.3 that cell division is usually a relatively short phase in a longer cell cycle in which the cell first grows and duplicates its contents, including the genomic DNA. Prokaryotes divide by binary fission to form two new individuals. The cells of more complex eukaryotes carry out one of two different forms of cell division.

■ What are the two forms of cell division carried out by complex eukaryotes?

☐ Mitotic cell division, which generates two daughter cells with the same number of chromosomes as the parent cell; and meiotic cell division, which produces haploid gametes (with half as many chromosomes as the parent cell).

During sexual reproduction in animals and plants, two haploid gametes fuse to form a diploid single-celled zygote. The zygote then develops by mitosis into the mature multicellular organism (Book 1, Section 4.3). After the initial growth and development phase, regulated mitotic cell division in the mature organism allows renewal and repair of some tissues throughout life, although, as you will see later, only certain types of cells in the tissues of mature multicellular organisms retain the ability to divide.

The primary concern of cell division is the accurate replication of the parent cell's genome and its equal partitioning between the daughter cells. A great deal of cellular infrastructure is involved in regulating cell division to ensure that the genomic information remains consistent between the generations of cells.

1.2.1 The bacterial cell division cycle

Cell division by binary fission in bacteria is tightly linked to cell size, and hence to cell growth and nutrient conditions. Bacteria grow bidirectionally from the middle outwards (Figure 1.1a and Book 1, Activity 4.1). New membrane and cell wall are synthesised and the cell cytoplasm increases in volume (shown as the paler-coloured regions of the cell in Figure 1.1a). The replication of the circular genomic DNA is initiated once the cell mass reaches a threshold level.

■ According to current understanding, how are the two genome copies physically segregated into the two daughter cells?

☐ The two DNA molecules attach to different points on the cell membrane which move apart as the cell elongates (Book 1, Activity 4.1).

Division takes place when the cell has roughly doubled in size (Figure 1.1a). Cytokinesis in bacteria involves the formation of a specialised structure called the Z ring (composed of proteins resembling some of the cytoskeletal proteins in eukaryotes, Book 1, Section 3.4.2), which attaches to the cell membrane and divides the cell in two by constriction, closely followed by the formation of a new cell wall (or septum) and separation of the daughter cells. Some species of bacteria undergo unequal division. For example, *Caulobacter crescentus* divides asymmetrically to form two unequal-sized daughter cells, one of which becomes a motile 'swarmer' cell and the other an immobile 'stalk' cell (Book 1, Figure 2.9d). Despite the apparent simplicity of the bacterial cell cycle, it is still quite poorly understood, and has only been studied in a few species, most notably *E. coli*.

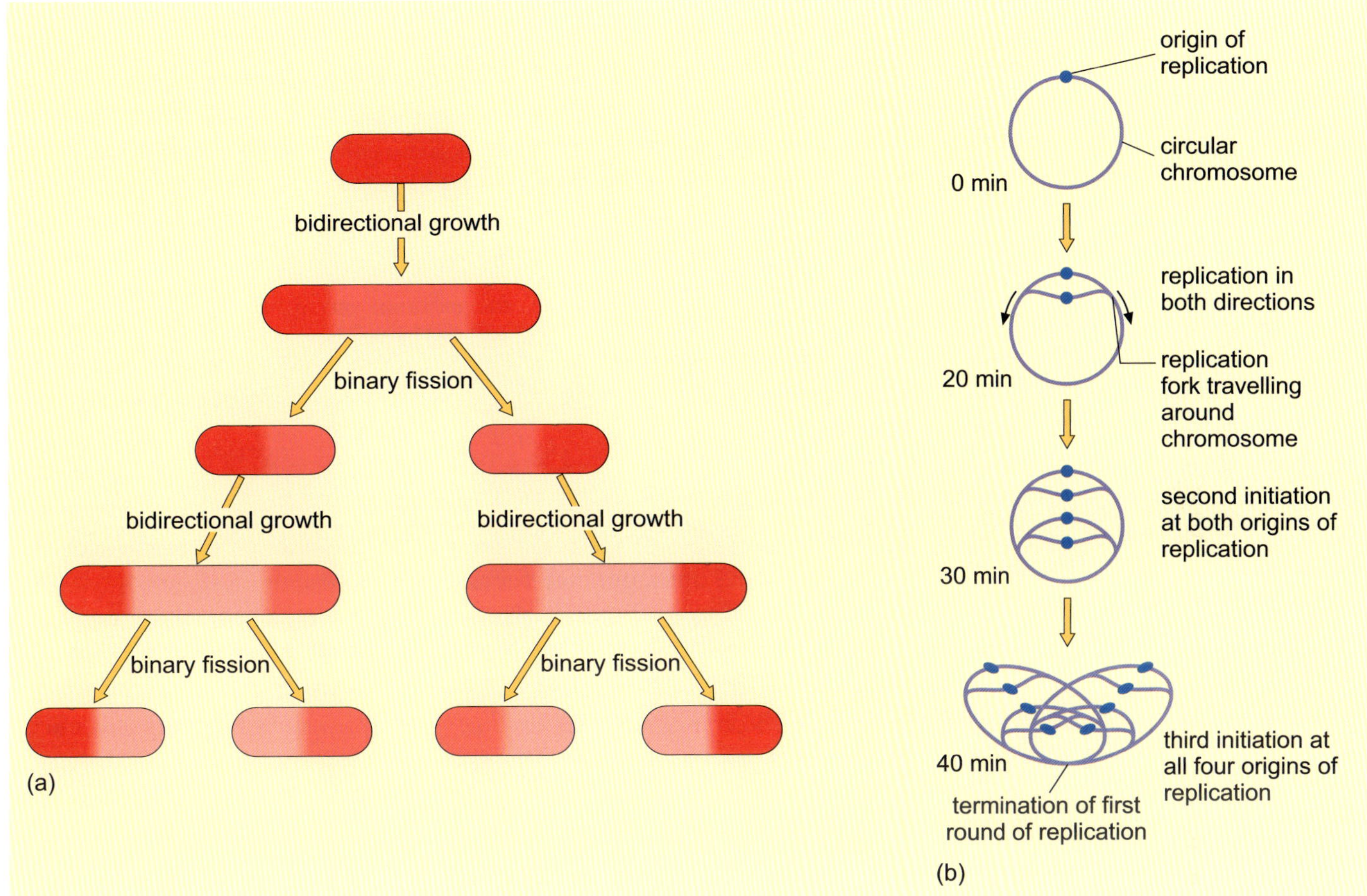

Figure 1.1 Binary fission and DNA replication in bacteria. (a) Diagram of the process whereby one cell divides to give two, these two each divide to give four, and so on. (b) Multiple initiations of DNA replication in bacteria that are growing with a doubling time much shorter than the time taken to replicate the complete chromosome (here, 40 minutes).

How is the bacterial cell cycle coupled to cell size? As in many other organisms, the initiation of DNA replication is a key event in cell cycle regulation. In most bacteria, initiation of DNA replication depends on the presence of a replication initiator protein called DnaA. DnaA is only active when it is bound to ATP. When cells are growing in a nutrient-rich medium, DnaA–ATP gradually accumulates as the cell grows in size. When cell mass reaches a certain threshold level, there are sufficient DnaA–ATP molecules to bind to the origin of replication on the bacterial DNA and unwind the DNA strands, allowing the replication complex access to the replication origin (Book 1, Section 5.3). Once DNA replication is initiated, hydrolysis of the DnaA-bound ATP to ADP inactivates all the DnaA molecules. DnaA–ATP must gradually accumulate again as the daughter cells grow. This ensures that replication of the bacterial chromosome is initiated only once for each cell division. As you will see throughout this section, the oscillating (rising and falling) activity of regulatory proteins like DnaA is a mechanism common to cell cycle regulation in all organisms.

In optimum, nutrient-rich conditions, *E. coli* can double in size in around 20 minutes, while the completion of one round of DNA replication requires 40 minutes. However, despite this discrepancy, replication can keep pace with cell growth and division because a second (and even third or fourth) round of DNA replication can be started before the first round is completed. Each new cell can receive DNA with two or more active replication forks and DNA replication proceeds continuously (Figure 1.1b), but each daughter cell eventually receives a single complete copy of the genome.

It is thought that various (little understood) signalling pathways allow bacterial cells to constantly 'sample' the external environment and transmit information about available nutrition and cell size directly to the cell cycle machinery. The time it takes for a bacterial population to double in number is known as the doubling time or the **generation time**.

■ Nutrient-rich conditions lead to a more rapid increase in cell size. What effect would this have on the generation time?

□ The generation time of the population would be shorter, because each cell would reach the optimum size for division more quickly.

Nutrient-poor conditions, on the other hand, reduce cell growth, increase generation time and curtail population growth.

The size of a cell population over time can be represented by a **growth curve** like the one shown in Figure 1.2. This type of growth curve is applicable not just to bacteria growing in a laboratory flask, but to any population of organisms growing in limiting conditions, even though they may not be reproducing by binary fission.

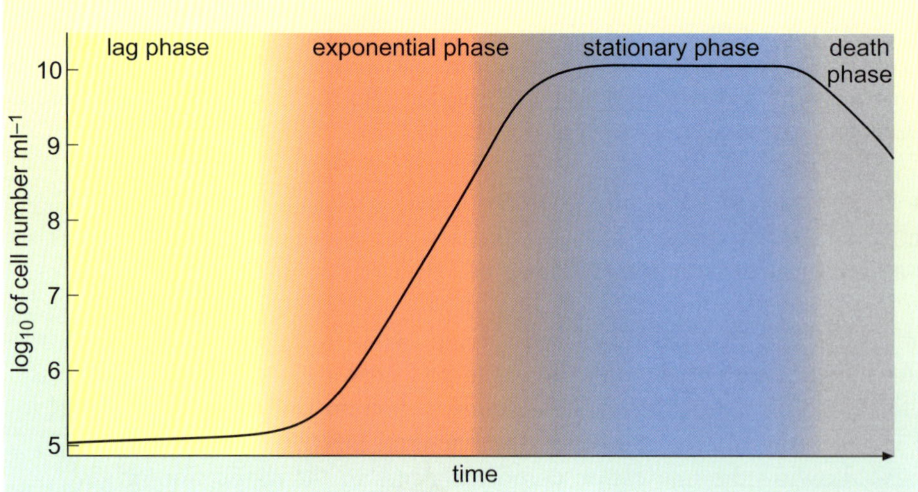

Figure 1.2 A graph showing a bacterial growth curve in a culture in which the cells are reproducing by binary fission. The lag, exponential, stationary and death phases are indicated (see text). Note that the *y*-axis is a logarithmic scale (see Box 1.1).

Box 1.1 Logarithms

A logarithm (or log) is the exponent or 'power' by which another value (the base) must be raised to obtain that number. For example, the logarithm of 1000 to base 10 ($\log_{10}$) is 3, because 1000 is 10 to the power 3 (or $10 \times 10 \times 10 = 10^3$). Thus, if the y-axis of the graph in Figure 1.2 represents $\log_{10}$ values, each increment on the scale is 10-fold greater than the previous one, i.e. the first increment represents 10^5 (100 000) cells, the next 10^6 (1 000 000) cells, the next 10^7 cells and so on. Plotting the logarithm (log) of the cell number is a convenient way of displaying data that covers a very large range of values. A logarithmic graph also makes it much easier to see the rate of change over time; in this case the growth rate of a bacterial culture.

When bacteria are initially introduced into a new environment, such as a culture medium containing a new carbon source, their growth is initially slow as they turn on genes encoding the enzymes required to metabolise the new nutrient. This is called the **lag phase** (Figure 1.2). Expression of the appropriate genes (such as the *lac* genes in the presence of lactose, Book 1, Section 6.3.3) then enables the bacteria to become optimally adapted to their new environment so that they grow and divide at the maximum rate possible for the conditions. This is called the **exponential phase**, which is characterised by doubling in cell numbers at equal time intervals; that is, one cell becomes two, two become four, four become eight, and so on.

For an exponentially dividing bacterial culture, the relationship between the initial number of bacterial cells in a population (N_o), the number of bacteria in a population at a given time later on (N_t), and the number of divisions those bacteria have undergone during that time (n) can be expressed by the following equation:

$$N_t = N_\mathrm{o} \times 2^n$$

Under optimal conditions, *E. coli* has a generation time of around 20 minutes. So, for example, starting with a hundred *E. coli* cells ($N_\mathrm{o} = 100$) that are allowed to grow exponentially for six hours, the population would divide three times during one hour, so six hours is enough time for 18 doublings ($n = 18$); therefore the number of bacteria after six hours (N_t) would be:

$$N_t = 100 \times 2^{18} = 26\ 214\ 400 \text{ or approximately } 2.6 \times 10^7 \text{ } E.\ coli \text{ cells.}$$

■ Starting with a culture containing 400 *E. coli* cells, how many cells would be present after two hours under optimum conditions?

□ The population would divide six times, so the number of bacteria present after two hours would be $400 \times 2^6 = 400 \times 64 = 25\ 600$ cells or approximately 2.6×10^4 *E. coli* cells.

■   As well as nutrient availability, what other environmental factors might affect generation time?

□   A variety of factors, including temperature and pH can affect the activity of enzymes (Book 2, Section 1.8.3), and therefore affect the rate of growth and generation time. You may also have thought of the effects of predators, pathogens (bacteriophage infections) or a build-up of toxic waste products in the cell.

You will learn how some bacterial populations have adapted to grow optimally in some extreme environments in Chapter 2 of this book.

Exponential growth cannot continue indefinitely in a closed bacterial culture. Increasing cell numbers will eventually start to deplete the available nutrient supply, and inhibitory waste products will begin to accumulate, so the growth rate will slow. This is known as the **stationary phase** (Figure 1.2).

Finally, when all the nutrients are used up, the culture contains mainly dead and dying cells and their waste products, and the number of cells begins to decline, a period known as the **death phase** (Figure 1.2).

■   Why do the cells die at this stage?

□   They may be unable to obtain enough precursors to sustain growth and cell maintenance. In this event, metabolic processes become so disrupted that life can no longer be sustained. The cells may also be poisoned by high levels of toxic molecules that they can no longer detoxify.

1.2.2   The eukaryotic cell division cycle

In Book 1, Chapter 4 you encountered the eukaryotic cell cycle; the ordered series of events that result in cell division by mitosis (or meiosis in germ line cells). To ensure that each step of the cell cycle occurs in the correct order (e.g. chromosome segregation should not occur before DNA replication has been completed), cells have evolved a complex regulatory system; a series of biochemical switches that control the timing of events.

In multicellular organisms the cell division cycle must also respond to extracellular signals. The rate of production of new cells must match the rate of cell loss. Old or damaged cells are normally eliminated by a process called *apoptosis*, or programmed cell death (reviewed in Section 1.3 of this chapter). Extracellular signals that maintain a balance between the processes of cell division and cell survival are therefore necessary to regulate cell numbers and maintain the shape and function of organs and tissues. Cancer cells have usually lost the ability to respond to extracellular regulatory signals, so uncontrolled cell proliferation and the evasion of programmed cell death are typical characteristics of cancer. If cancer cells remain in a small discrete group within their tissue, the cancer is often benign; however, malignancy occurs if the proliferating cells acquire the ability to invade and disrupt surrounding tissue.

Recapping the stages of the cell cycle

You will recall from Book 1, Section 4.3 that the eukaryotic cell cycle consists of two main phases: M phase, during which the cell actually divides; and the much longer period between M phases, known as interphase (Figure 1.3). Interphase can be subdivided into three phases, starting with an initial period of cell growth, called G1 phase. G1 is followed by S phase, during which the nuclear DNA is replicated. After S phase is G2 phase, when the cell grows and prepares for division in M phase. If the signals that trigger cell proliferation are absent, cells may exit the cycle at G1 and enter a resting or quiescent state, known as G0 (G zero).

The duration of the cell cycle varies in different cell types. For example, the single-celled yeast *Saccharomyces cerevisiae* has a cell cycle lasting only about 90 minutes, while mammalian cells growing in laboratory cultures typically have a cell cycle of approximately 20 hours. In multicellular organisms, many fully differentiated cells (such as neurons) have completely withdrawn from the cell cycle and do not divide.

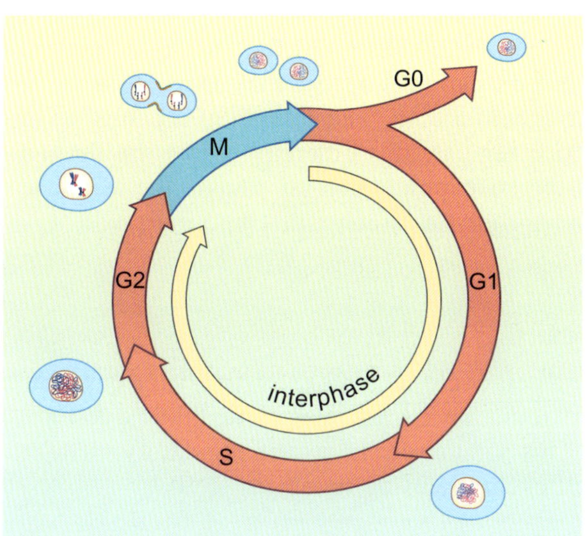

Figure 1.3   Diagram of the cell cycle. Cells actually divide during M phase, when they undergo mitosis and cytokinesis. The period between successive M phases is known as interphase. During the first part of interphase (G1), cells undergo a period of growth. This is followed by DNA replication (S), and a further period of growth (G2). Cells can temporarily or permanently withdraw from the cell cycle at G1 and enter a period of rest or quiescence, known as G0.

M phase includes two overlapping processes. The first is mitosis in which the two copies of each chromosome, termed sister chromatids, which until this point have been closely associated, are separated and segregated into two nuclei. This is followed by cytokinesis, the process in which the cell cytoplasm divides to form two completely separate daughter cells. M phase usually lasts only about an hour in a typical mammalian cell cycle of total duration 20 hours.

The two gap phases, G1 and G2, allow the cell time to increase its mass and replicate organelles and proteins before division. Exceptions are the rapid early divisions (known as cleavages) of animal zygotes, which are very large in size relative to somatic cells and contain sufficient resources for several divisions to occur without growth, so that cells of successively smaller size are formed with each division. The only time needed between divisions of these cells is thus the time needed for DNA replication, so their cell cycle consists of short S and M phases with no gap phases.

An additional role of the gap phases is to allow the cell cycle regulatory system time to check that everything is ready for the next stage, and to carry out any necessary repairs.

■ What might need to be 'checked' during the G1 and G2 phases of the cell cycle?

□ During G1, the readiness of the cell for DNA replication, including the condition of the DNA and the amounts of other cell components, requires checking. During G2, the successful completion of DNA replication needs to be confirmed before the cell goes ahead and divides.

There are in fact several so-called **cell cycle checkpoints** which ensure that each phase of the cell cycle has been accurately completed before the cell progresses to the next phase. These checkpoints ensure that damaged or abnormal cells are prevented from proliferating. The following sections describe the molecular switches that regulate this ordered progression through each stage of the cell cycle, and the checkpoints that coordinate the response to external signals and prevent inappropriate or aberrant (abnormal) cell division.

1.2.3 Cell cycle regulators

The components of the eukaryotic cell cycle regulatory system vary between different organisms, but there are some basic principles of regulation that are common to all eukaryotic cells. This section considers the regulatory switches that trigger each phase of the cell cycle, focusing on animal cells.

Cell cycle kinases

Like many processes in the cell, the progression through the cell cycle is regulated by the sequential activation and inactivation of a number of regulatory proteins that initiate key steps, such as the initiation of DNA replication or chromosome segregation.

■ What two main types of reversible modification do cells use to activate and inactivate enzymes and other proteins?

□ Binding of an activator or repressor molecule (allosteric regulation), and reversible covalent modification of the protein itself (e.g. phosphorylation of specific amino acid residues) (Book 2, Section 1.8.5).

The key regulators of progression through the cell cycle in all eukaryotic cells are a family of protein kinases known as **cyclin-dependent protein kinases (Cdks)**. At the appropriate points in the cell cycle, these Cdks phosphorylate and thereby activate particular target proteins that are necessary to complete each stage of the cell cycle. The Cdks are constitutively expressed, so they are always present in the cell, but they are only active during particular phases of the cell cycle and are then promptly deactivated again until the next cycle. Cdks are activated at the appropriate time by their interaction with another type of protein called a **cyclin**. Cyclins have no enzyme activity of their own, but their expression level in the cell rises and falls during the cell cycle. When their concentration is high, cyclins bind to and activate their partner

Cdk (hence the name, cyclin-dependent protein kinases). Thus, the oscillating levels of cyclins ensure that Cdks are only activated at the appropriate time in the cell cycle to trigger events by phosphorylating and activating their target proteins.

For example, the M phase cyclin (M cyclin) that helps to drive cells into M phase (see below) starts to be synthesised in interphase, and slowly accumulates until M phase begins. M cyclin binds to and activates the appropriate Cdk and triggers the processes of mitosis. Once this is accomplished, the cyclin rapidly disappears again (Figure 1.4a).

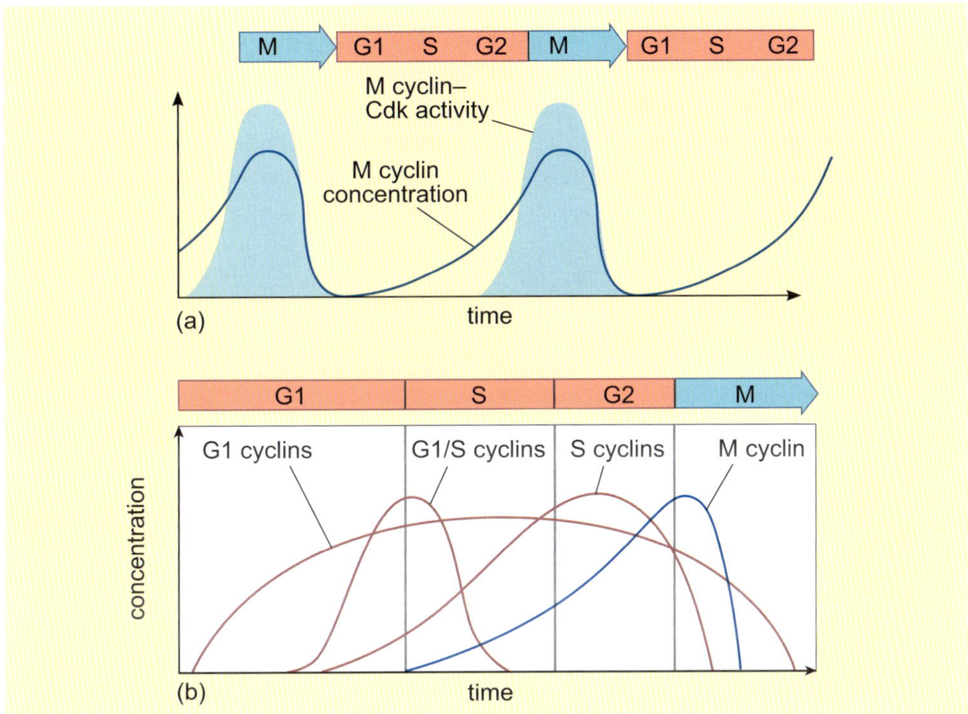

Figure 1.4 The cyclic increase and decrease in the levels of cyclins. (a) M cyclin is synthesised during interphase, reaching a peak level at the start of mitosis, and is then rapidly degraded. Maximal M cyclin–Cdk activity also requires phosphorylation of the complex. (b) A graph indicating the general changes in concentration of different classes of cyclin proteins during the stages of the cell cycle in human cells (the sequence is similar in other organisms but the periodicity and numbers of cyclins vary).

■ What mechanism would ensure that cyclin protein levels fall very rapidly once they have performed their role at a particular stage of the cycle?

□ They are rapidly degraded by proteases.

The cyclins are targeted for destruction by the addition of chains of ubiquitin molecules which directs the proteins to the proteolytic enzymes of the proteasome (Book 2, Section 1.5.4). The Cdk partner of the cyclin is thereby released and returns to its inactive state until the cyclin accumulates again in the next cell cycle.

While the gradual rise and fall of cyclin concentrations is crucial in coordinating the activity of cyclin–Cdks, it is noticeable that the complexes become active quite abruptly, for example M cyclin–Cdk is activated suddenly at the end of interphase (Figure 1.4a). What triggers this? Full activation of cyclin–Cdk complex requires not only cyclin binding, but also the phosphorylation or dephosphorylation (by specific kinases and phosphatases) of particular amino acid side chains close to the active site of the Cdk protein. This reversible covalent modification of the Cdk partner provides the final finely tuned coordination of cyclin–Cdk activation.

In most eukaryotes there are several different types of Cdk, which are involved in the regulation of different stages of the cell cycle. Notable exceptions are the yeasts, which have a single type of Cdk. However, both yeast and mammalian cells have several types of cyclin that are synthesised at different stages of the cell cycle. The regulatory cyclin subunit to which the Cdk binds determines the specificity of the Cdk; in other words, it determines which target proteins the cyclin–Cdk will phosphorylate and activate. The presence of different cyclins at different stages of the cell cycle therefore enables the single type of Cdk in yeast to control different phases of the cell cycle.

The activation of Cdks by binding to cyclins at particular phases thus drives the orderly progression through the events of the cell cycle. Although their exact function differs between organisms, cyclins can be grouped into four general classes that relate to the phase of the cell cycle that they regulate: G1 cyclins promote entry into the cell cycle; G1/S cyclins trigger the transition from G1 to S; S cyclins promote DNA replication (although their level remains high well into G2 phase because they also promote some of the events in early mitosis); and M cyclins trigger the main events of mitosis (Figure 1.4b).

1.2.4 G1 cyclin-Cdks promote entry into the cell cycle

Most of the differentiated cells in a mature multicellular organism have withdrawn from the cell cycle and exist in G0, the quiescent state where they express no Cdks or cyclins and don't proliferate. In cells that do retain the ability to enter the cycle, the decision to do so usually depends on extracellular signals which stimulate the synthesis of the G1 cyclins.

- ■ What type of extracellular signals might be received and transduced into the cell to stimulate G1 cyclin expression?

- ☐ The binding of growth factors to cell surface receptors (Book 2, Section 4.2).

The point in G1 at which a cell becomes committed to the cell cycle is referred to as the **restriction point** (or sometimes just 'start'). Once a cell passes the restriction point, it is committed to completing the rest of the cell cycle. The restriction point is therefore the first of several important checkpoints in the cycle and it monitors several factors, including nutrient availability, cell size and the presence of growth factors (Figure 1.5). You will learn more about the role of different types of growth factor in managing cell numbers later on in this chapter.

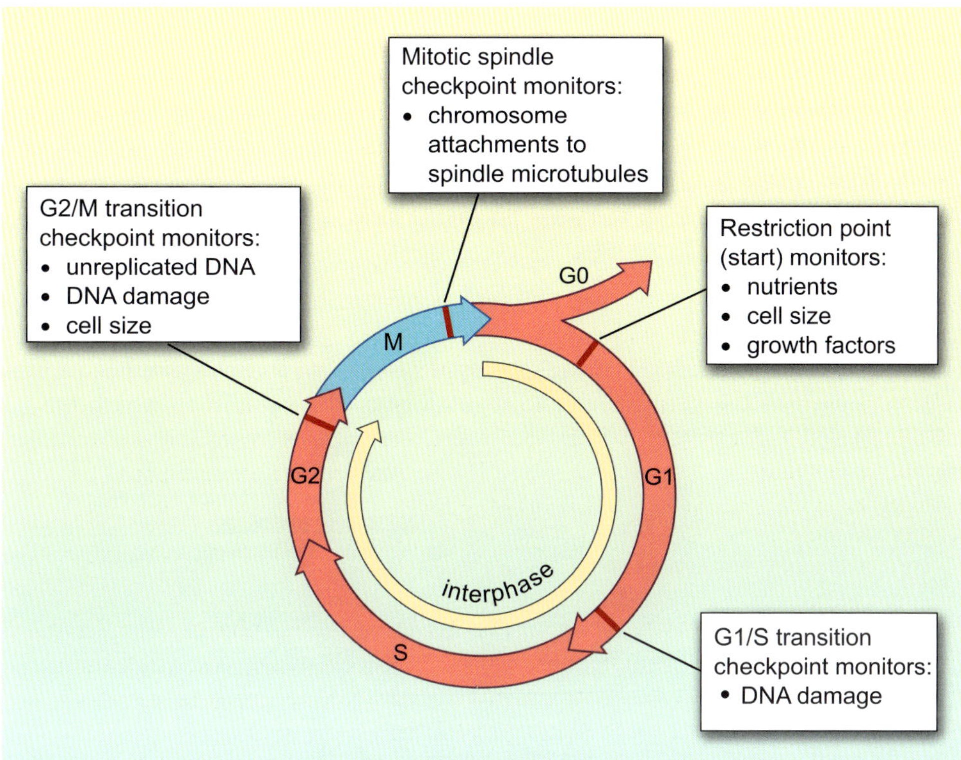

Figure 1.5 Several checkpoints monitor the progress of different events in the cell cycle and can arrest cells at G1, G2 or M phase if the cell is not ready to progress.

When the conditions are appropriate, the cell is stimulated to synthesise G1 cyclins. Note that, unlike the other cyclins shown in Figure 1.4b, G1 cyclin levels increase gradually throughout the cycle in response to cell growth and extracellular signals. As their levels increase, active G1 cyclin–Cdk complexes are formed which phosphorylate a protein known as **retinoblastoma (Rb) protein**, which is present in the cell nucleus. While Rb protein remains in its *unphosphorylated* state, it binds to and inhibits the actions of several transcription factors that drive the expression of key proteins required for cell division, including those required for DNA replication. The synthesis of these proteins is thus prevented (Figure 1.6a). However, once the Rb protein is phosphorylated by G1 cyclin–Cdk complexes, it releases these transcription factors, so the proteins needed for DNA replication are synthesised and the cell can progress through the cell cycle (Figure 1.6b). Quiescent cells that have withdrawn from the cell cycle (entered G0) have effectively 'dismantled' the cell cycle machinery, and have stopped making cyclins and Cdks. Thus Rb is not phosphorylated in these cells, and transcription of the proteins needed for DNA replication cannot occur. Rb thus acts like a brake that prevents cells from entering the cycle until conditions are appropriate.

The *Rb* gene is mutated in many types of cancers and the Rb protein was first identified because it is either absent or abnormal in the cells of retinoblastoma, a rare form of tumour of the eye that occurs in children. Rb is an example of a so-called **tumour suppressor**; if it is absent or defective, its braking effect on the cell cycle is lost, which can result in uncontrolled cell proliferation.

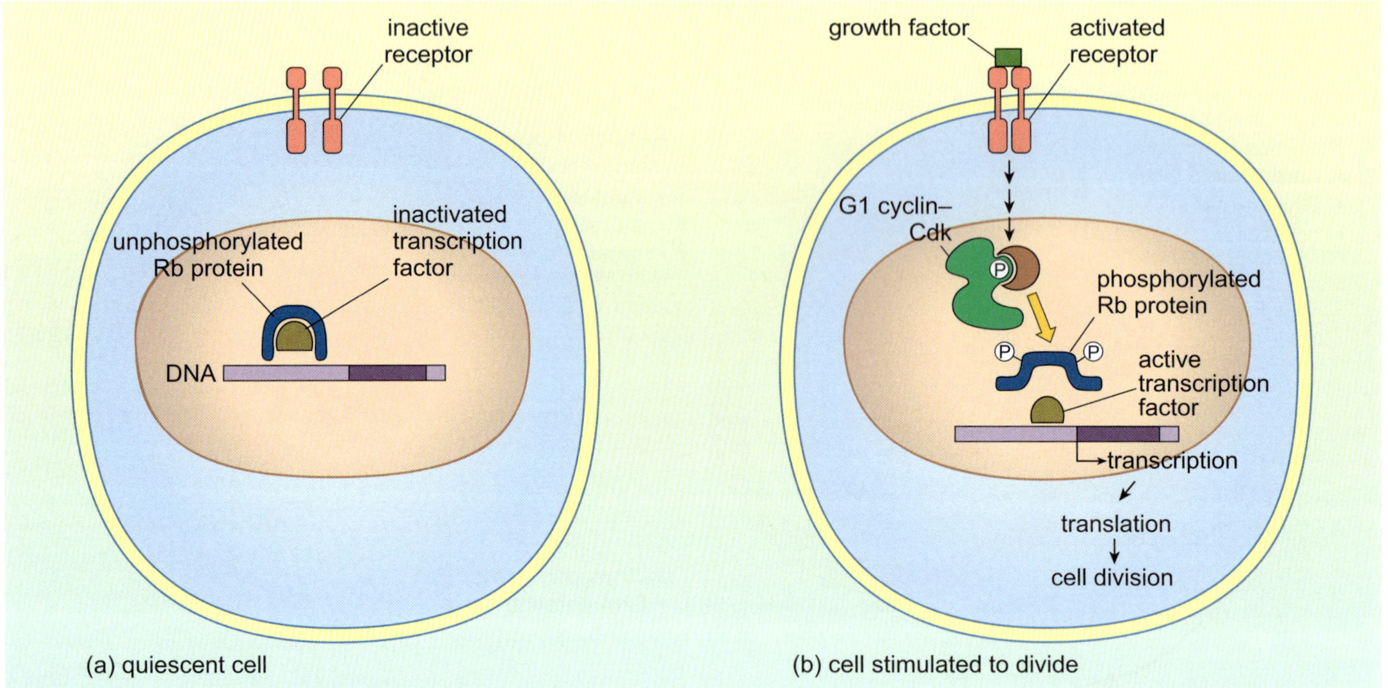

Figure 1.6 G1 cyclin–Cdks phosphorylate the Rb (retinoblastoma) protein thereby allowing cells to enter the cell cycle. (a) If growth factors are absent, the unphosphorylated Rb protein inactivates transcription factors that are needed for transcription of genes encoding proteins needed for DNA replication. (b) When growth factors are present, they activate signalling pathways that result in the synthesis and activation (by phosphorylation) of the G1 cyclin–Cdks. The activated G1 cyclin–Cdks in turn phosphorylate Rb, which then releases the transcription factors that stimulate transcription of the genes needed for cell division.

1.2.5 G1/S and S cyclin-Cdks promote DNA replication

Among the cell cycle proteins synthesised as a result of Rb phosphorylation in G1 are the next set of cyclins in the sequence – the G1/S cyclins and S cyclins.

- What molecules would you expect to be phosphorylated and activated by cyclin–Cdks at the transition from G1 to S phase?

- ☐ Molecules that form part of the DNA replication machinery.

G1/S cyclin–Cdks are active at the start of S phase (Figure 1.4b). They promote assembly of the DNA replication complexes at origins of replication on the DNA (Book 1, Section 5.3). They also trigger duplication of the centrosome, the microtubule organising centre in animal cells that will be required to form the mitotic spindle (Book 1, Section 3.4.2). S cyclins are present throughout S phase and they stimulate the ongoing replication of the DNA.

Checkpoints that monitor DNA integrity

It would obviously be problematic if cells were able to continue through the cell cycle without completing the replication of their DNA, or indeed if their DNA was damaged. A key factor in determining whether a cell will complete the cycle therefore is the state of its DNA. If stopped or 'stalled' DNA replication forks are detected, a **DNA replication checkpoint** is activated. This prevents the activation of M cyclin–Cdk complexes (see below) so that the cell is halted at the transition from G2 to M (Figure 1.5), unless it can resolve the problem.

In addition, damage to DNA is monitored throughout the cycle by **DNA damage checkpoints** that can halt the cycle at the transitions from G1 to S or G2 to M by inhibiting cyclin–Cdk complexes. A protein called p53, often referred to as the 'guardian of the genome', has a key role in these two DNA damage checkpoints. When DNA damage is detected in the cell, the p53 protein is stabilised, and activated by phosphorylation (Figure 1.7).

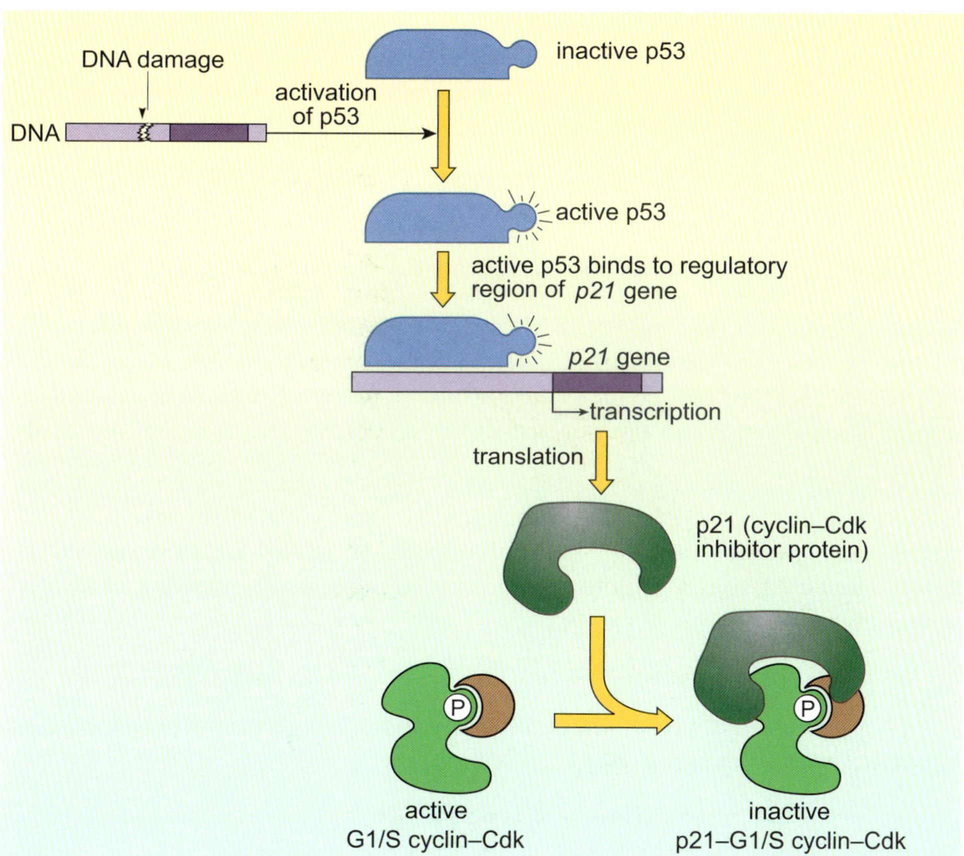

Figure 1.7   The role of p53 and p21 in the arrest of the cell cycle when DNA is damaged. DNA damage results in activation of the transcription factor p53, which stimulates transcription of the *p21* gene. p21 protein binds to cyclin–Cdk complexes and thereby inactivates them. This figure shows inactivation of G1/S cyclin–Cdk, which leads to arrest of the cycle at the G1 to S transition (Figure 1.5).

Activated p53 does many things; one is to stimulate the transcription of the gene that encodes a protein known as p21. p21 binds to and inactivates cyclin–Cdk complexes and thus stops cell cycle progression (at the transitions from G1 to S or G2 to M, Figure 1.5). This arrest allows time for DNA repair before the cell is allowed to continue through the cell cycle. However, if the damage is too severe, the cell may die. This death is not 'accidental', but a form of cell suicide called apoptosis that you will learn more about later in this chapter. This control mechanism is very important; mutations in the *p53* gene are involved in the formation of many cancers because they allow cells with damaged DNA to continue dividing unchecked. Like the Rb protein, p53 is a tumour suppressor because it has a role in preventing inappropriate cell proliferation.

1.2.6 M cyclin-Cdks initiate mitosis

Perhaps the most dramatic phase of the cell cycle is M phase, during which the cell completely reorganises its components in order to be able to separate into two daughter cells. This extremely complex process is triggered by the accumulation of M cyclin and the abrupt activation of M cyclin–Cdk complexes (Figure 1.4a), which initiates a whole range of early events in mitosis.

■   What events occur during the first stage of mitosis (prophase)?

□   The nuclear envelope breaks down, the chromosomes condense and the mitotic spindle forms (Book 1, Section 4.3.1 and Activity 4.1).

As prophase begins, the activated M cyclin–Cdk phosphorylates protein complexes called *condensins* that assemble onto the chromatin and promote chromosome coiling and condensation (Book 1, Figure 3.12). The condensed mitotic chromosomes can be more easily segregated within the confines of the cell.

At the same time, M cyclin–Cdk promotes the separation of the two centrosomes (Figure 1.8) which were duplicated in interphase as described above. The centrosomes migrate along microtubules to the opposite ends or 'poles' of the cell.

■   How are structures like centrosomes moved along microtubules?

□   They are carried along microtubules by cytoplasmic motor proteins (kinesins and dyneins, Book 2, Section 5.2.2).

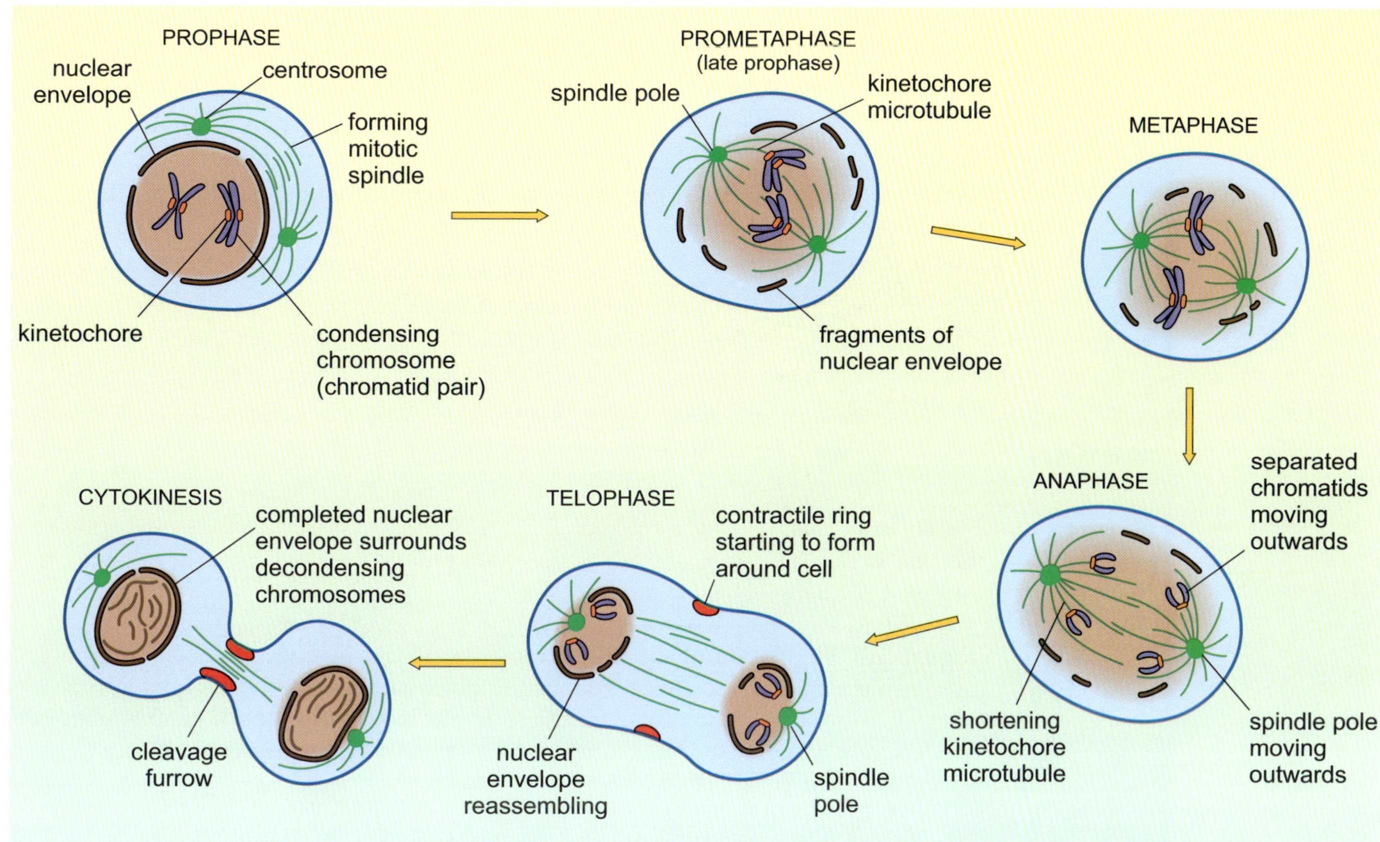

Figure 1.8 Mitosis and cytokinesis. In prophase, the chromatin condenses and sister chromatids, resulting from the duplication of the chromosomes during S phase, are visible. The nucleolus disappears and the mitotic spindle begins to form between the duplicated centrosomes. In prometaphase (late prophase), the nuclear envelope breaks down and chromatid pairs bind to the spindle via kinetochores. During metaphase, these chromatids line up along the equator of the mitotic spindle. Sister chromatids are pulled to opposite poles of the cell during anaphase, and daughter nuclei are formed during telophase. Division of cytoplasmic contents and physical separation of daughter cells are achieved during cytokinesis.

The mitotic spindle, the structure that is critical for the separation of the chromatids during mitosis, begins to form as microtubules start to extend from each centrosome (Figure 1.9). The minus end of each spindle microtubule is stabilised by association with the centrosome, but the plus ends are highly dynamic and undergo rapid shrinking and growing (Book 2, Section 5.2.1). The M cyclin–Cdks phosphorylate and activate **microtubule associated proteins** which bind to the plus ends of the microtubules and regulate their stability. The microtubule associated proteins cross-link microtubules as they overlap in the centre of the cell, thereby stabilising the plus ends and preventing depolymerisation (Figure 1.9). Eventually the spindle-shaped microtubule structure stretches across the centre of the cell.

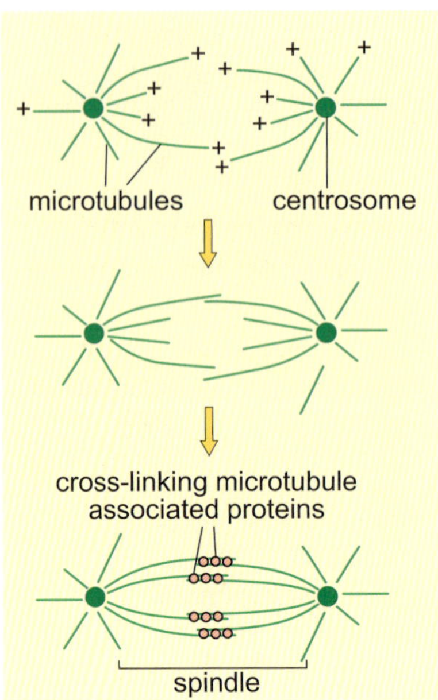

Figure 1.9 Formation of the mitotic spindle. M cyclin–Cdk promotes migration of the duplicated centrosomes to opposite poles and also activates microtubule associated proteins which regulate microtubule growth from each centrosome. Microtubule associated proteins also cross-link overlapping microtubules, stabilising their plus (+) ends.

At the end of prophase (in a stage often referred to separately as prometaphase, Figure 1.8), the M cyclin–Cdk phosphorylates lamins (the intermediate filament proteins on the inner surface of the nuclear membrane, Book 1, Section 3.4.3). This promotes the disassembly of the nuclear lamina and the breakdown of the nuclear membrane. The replicated chromosomes thereby released from the nucleus can now be captured by the plus ends of microtubules extending from the centrosome. Spindle microtubules attach to a protein complex called the **kinetochore** which forms on each chromosome during prophase. The kinetochore proteins assemble at the centromere, shown as a constricted region on the paired chromatids in Figure 1.10, such that each chromatid has its own kinetochore complex. Microtubules that have 'captured' a kinetochore are called **kinetochore microtubules**. In pairs of sister chromatids, the kinetochores face in opposite directions and bind to microtubules projecting from opposite poles of the cell, thus ensuring that the two chromatids will be segregated to opposite ends of the cell.

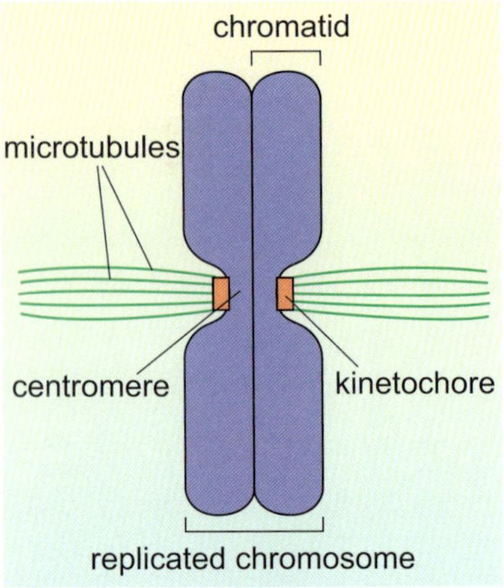

At metaphase (Figure 1.8) the chromosomes, with their two chromatids attached to microtubules from opposite poles, have lined up on an imaginary plane known as the **metaphase plate**, which is located at the equator (centre) of the spindle, perpendicular to the long axis of the spindle.

The mitotic spindle checkpoint

At metaphase, the **mitotic spindle checkpoint** (Figure 1.5) is in place to prevent chromosome segregation from occurring before all the chromosomes have correctly lined up at the metaphase plate. If, for example, a kinetochore has become inappropriately attached to a microtubule extending from the

Figure 1.10 Arrangement of sister chromatids on microtubules of the spindle and association of kinetochores with the centromere.

wrong pole, the kinetochore will detach and reattach to another microtubule. Mitosis cannot continue until every chromosome is correctly oriented.

- What might be the consequence of mitosis in a cell with a defective mitotic spindle checkpoint?

- If some chromatids attach to microtubules from the wrong pole, the daughter cells may acquire an uneven complement of chromosomes; one may obtain too many and the other too few.

The cells would thus have an abnormal number of chromosomes, which is known as **aneuploidy**. This can have disastrous consequences for gene expression and the function of the cell. Aneuploidy is a common cause of miscarriages and birth defects and is also a characteristic of most cancerous cells.

The checkpoints that detect the completion of DNA replication, DNA damage and chromosome orientation and attachment thus all ensure that the daughter cells acquire intact copies of the parent cell's genome. A cell that halts at one of these checkpoints and fails to correct the problem is likely to trigger its own death, thus avoiding the danger of producing aberrant or even cancerous daughter cells.

1.2.7   Destruction of M cyclins triggers the completion of mitosis and cell division

Once the cell has reached metaphase with all of the chromatids correctly attached to the spindle, M cyclin rapidly disappears, so M cyclin–Cdk activity is lost.

- What mechanism triggers the disappearance of M cyclins?

- The cyclin molecules become 'tagged' with molecules of ubiquitin, which identifies them as targets for degradation by the proteasome (Section 1.2.3).

The ubiquitination of M cyclins is carried out by a ubiquitin ligase called **anaphase promoting complex (APC)**, which is itself activated by the M cyclin–Cdk complex, so the cyclins in effect trigger their own destruction once their job is done.

The APC not only tags M cyclin for destruction, but also breaks down proteins that hold the duplicated chromatids together, triggering chromatid separation (Activity 4.1).

- What is the name of the phase of mitosis during which chromatids separate to the poles of the cell?

- Anaphase.

Hence the name 'anaphase promoting complex' for the complex that brings about inactivation of M cyclin–Cdk.

During anaphase, each chromatid moves towards its associated spindle pole (Figure 1.8). This movement is accomplished by two types of microtubule-dependent movement. Firstly, the kinetochore microtubules start to disassemble at their plus ends, so effectively, the microtubules become shorter and their cargo (the chromatids) moves towards the pole. Secondly, motor proteins attached to cross-linked overlapping microtubules (which have not captured a chromatid) slide the microtubules apart. This has the effect of pushing the spindle poles away from each other.

The last stage of mitosis is telophase (Figure 1.8) in which a nuclear envelope assembles around each group of chromosomes, forming two nuclei. The chromosomes decondense and the mitotic spindle disassembles once mitosis is complete.

The final stage of the cell cycle is cytokinesis (Figure 1.8) in which the cytoplasm and organelles of the dividing cell are partitioned to give two daughter cells. Cytokinesis in animal cells involves a contractile ring composed of overlapping filaments of actin and myosin which assemble around the equator of the cell. The contractile ring is linked to membrane-associated proteins on the cytoplasmic surface of the cell membrane, so as it constricts (the actin filaments slide over the myosin filaments by a mechanism similar to that observed in muscle contraction, Book 2, Section 5.3), a cleavage furrow forms which pinches inwards until the daughter cells are separated.

1.2.8 Control of cell number and size

So far, this chapter has considered the control of the cell cycle in cells that are constantly proliferating. In multicellular organisms, this type of continuous rapid cell division occurs during early development, but most cells eventually differentiate to form mature tissues and organs in adult organism and stop dividing, so that cell proliferation becomes strictly controlled to match need. What are the mechanisms that operate to prevent the inappropriate proliferation of cells?

As described earlier, entry into the cell cycle is initiated by extracellular signals that usually originate from a cell's neighbours.

■ What are the two main strategies that neighbouring cells use to signal to each other?

☐ Physical contact between molecules on their surface (contact-dependent signalling) and secretion of soluble factors that bind to receptors on neighbouring cells (Book 2, Section 4.2).

The collective term 'growth factor' is applied (often rather misleadingly) to a whole range of extracellular signalling molecules with varying abilities to promote cell survival, growth, proliferation, differentiation or even cell death in different cell types. Most growth factors are small polypeptides secreted by cells to target nearby cells that carry appropriate cell surface receptors. Binding of growth factors to their receptors activates intracellular signalling cascades that influence gene expression (Book 2, Chapter 4).

There are several families of growth factor, each consisting of several related polypeptides. Some factors are very specific in their action, selectively promoting the proliferation or differentiation of a particular cell type, whereas others are active on a range of cell types and may have multiple effects. Examples of some growth factors are shown in Table 1.1.

Many growth factors act as **mitogens** which stimulate cell division by overcoming inhibitory controls, such as the Rb protein (Section 1.2.4), that prevent progression through the cell cycle. Mitogens promote proliferation by activating intracellular signalling pathways that increase the synthesis of G1 cyclins. The activated G1 cyclin–Cdk complexes then phosphorylate Rb, relieving its inhibition of transcription and thereby allowing the cell to progress through the cell cycle (Figure 1.6).

Other growth factors promote an increase in cell mass and size, which is a prerequisite for cell division (otherwise dividing cells would become progressively smaller). In single-celled organisms, growth depends on environmental conditions, particularly nutrient availability, but in animal cells it also requires specific growth factors that stimulate intracellular signalling pathways and increase the production of proteins and cell components. Many cells (e.g. neurons) can continue to grow once they have differentiated and stopped dividing.

Table 1.1 Some examples of growth factors and their actions (this list is not exhaustive). In many cases, growth factors have a variety of effects, most of which are still being characterised.

Growth factor	Examples of actions
epidermal growth factor (EGF)	promotes proliferation, differentiation and survival of epithelial and other cell types
erythropoietin	promotes proliferation and differentiation of red blood cell precursors
fibroblast growth factor (FGF)	promotes proliferation and differentiation of a wide variety of cell types; essential in vertebrate and invertebrate development
insulin-like growth factor (IGF)	promotes survival and proliferation of many cell types; essential in early vertebrate development
nerve growth factor (NGF)	promotes survival and axon growth of some types of neurons
platelet-derived growth factor (PDGF)	promotes proliferation of connective tissue and smooth muscle cells; has a role in development, and in angiogenesis (blood vessel formation)
transforming growth factor β (TGFβ)	controls proliferation in most cell types by promoting differentiation and apoptosis

A third class, survival factors, promote cell survival, largely by suppressing a type of cell death called apoptosis, about which you will learn more in the next section. Most cells need to continuously receive signals from other cells in order to survive, and without them they may activate a 'suicide' programme. Survival factors usually activate intracellular signalling pathways that suppress apoptosis. An example of this is the regulation of nerve cell numbers in the developing nervous system, which you will encounter in the next section.

Tumour suppressors and oncogenes in cancer

A common feature of cancer cells is that they have acquired mutations in key cell cycle proteins. Most cancer cells have mutations that inactivate one or more tumour suppressors, such Rb or p53. This enables them to bypass cell cycle checkpoints and evade programmed cell death.

In addition, cancer cells commonly have gain of function mutations (Book 1 Section 5.6.2) that activate, or increase the expression of, proteins that *stimulate* cell division. Because of their potential to cause uncontrolled proliferation and contribute to the development of cancer, such mutated genes are called **oncogenes** (from the Greek *onkos*, meaning protuberance or swelling). The normal gene that, when mutated, can give rise to an oncogene, is known as a **proto-oncogene**, of which there are many types.

- Suggest some general types of protein that might be encoded by proto-oncogenes.

- Proto-oncogenes encode proteins that stimulate cell proliferation. Examples might include growth factors that stimulate proliferation, the receptors that these factors bind to, components of the intracellular signalling pathways (e.g. Ras protein, Book 2, Section 4.7.1) or transcription factors.

The gradual accumulation over time of multiple oncogene and tumour suppressor mutations in cancer cells usually correlates with increasing aggressiveness of a tumour. This is why cancer is most prevalent in older people. Researchers have therefore focused on these genes and their protein products in the search for anti-cancer treatments.

Summary of Sections 1.1 and 1.2

- Bacteria must adapt rapidly to changing environmental conditions, and their rate of proliferation is linked to cell size and nutrient availability. A typical bacterial growth curve includes four phases: lag phase, exponential phase, stationary phase and death phase.

- Regulation of the eukaryotic cell cycle in multicellular organisms requires growth signals to match proliferation to the requirement for new cells. The cycle consists of four main phases, known as G1, S, G2 and M. Cells can leave the cycle and enter a quiescent state, known as G0.

- The ordered sequence of molecular events required for progression through the eukaryotic cell cycle is controlled by protein complexes of cyclins and cyclin-dependent kinases (Cdks). The activity of these complexes changes at different stages of the cell cycle as cyclin levels rise and fall. Phosphorylation and dephosphorylation of cyclin–Cdk complexes adds a further level of fine control.

- Molecular checkpoints operate at several different stages in the cell cycle and prevent progression unless the cell is 'ready': for example, the DNA replication checkpoint prevents the cell from advancing to mitosis until all the DNA has been replicated.

- Rb protein is an example of tumour suppressor, a molecule that acts as a brake on the cycle. It binds to the transcription factors required for transcription of the genes needed for DNA replication. Upon phosphorylation by activated cyclin–Cdks, Rb protein releases these transcription factors so cell division can proceed.

- Another tumour suppressor, the transcription factor p53 plays a role in preventing progression through the cycle when DNA is damaged. When p53 is activated it promotes expression of the cyclin–Cdk inhibitor protein p21.

- Many cells that are in G0 can be stimulated to divide by growth factors produced by other cells. Polypeptide growth factors bind to cell surface receptors and activate intracellular signalling cascades, resulting in stimulation of the expression of the genes encoding cyclins.

- Genes that have products that normally act to suppress cell proliferation (including Rb and p53) are called tumour suppressor genes. Loss or mutation of tumour suppressor genes can contribute to uncontrolled cell proliferation. Most cancers have accumulated inactivating mutations in tumour suppressor genes and also activating mutations in oncogenes; genes whose products normally stimulate proliferation because they are involved in a pathway that transduces the signal from a stimulatory growth factor to regulate the expression of genes involved in the control of cell division.

1.3   Cell death

The previous section has already mentioned some situations where cell 'suicide' is evoked as a mechanism to remove cells that are damaged, or which are surplus to requirements. In fact, cell death plays an important part in normal development and in the maintenance of body tissues. In a mature multicellular organism the number of cells is tightly regulated, not only by control of the rate of cell division, but also by control of the rate of cell death.

An abnormally high level of cell death is a feature of several diseases, notably neurodegenerative diseases, such as Parkinson's disease and motor neuron disease. In contrast, cancer cells lack the normal controls on cell proliferation, and often evade naturally occurring cell death. From a therapeutic point of view, techniques for controlling natural cell death would clearly be of value in these situations.

The fate of an individual cell is determined by intricate molecular networks that achieve a balance between cell proliferation and cell death. This section begins with a description of two different types of cell death, again focusing on examples in animals (although both processes also occur in plants), before going on to outline the complex mechanisms that trigger cell death.

1.3.1   Types of cell death

There are two main ways in which cells can die. One is the typical response of cells to injury in the form of harmful reagents, infections or wounds. This type of death is known as **necrosis**. In necrosis, the damaged cell swells because the cell membrane fails to control the passage of ions and water, and finally lyses (bursts), releasing its contents (Figure 1.11a), which in vertebrates can stimulate a potentially damaging inflammatory response.

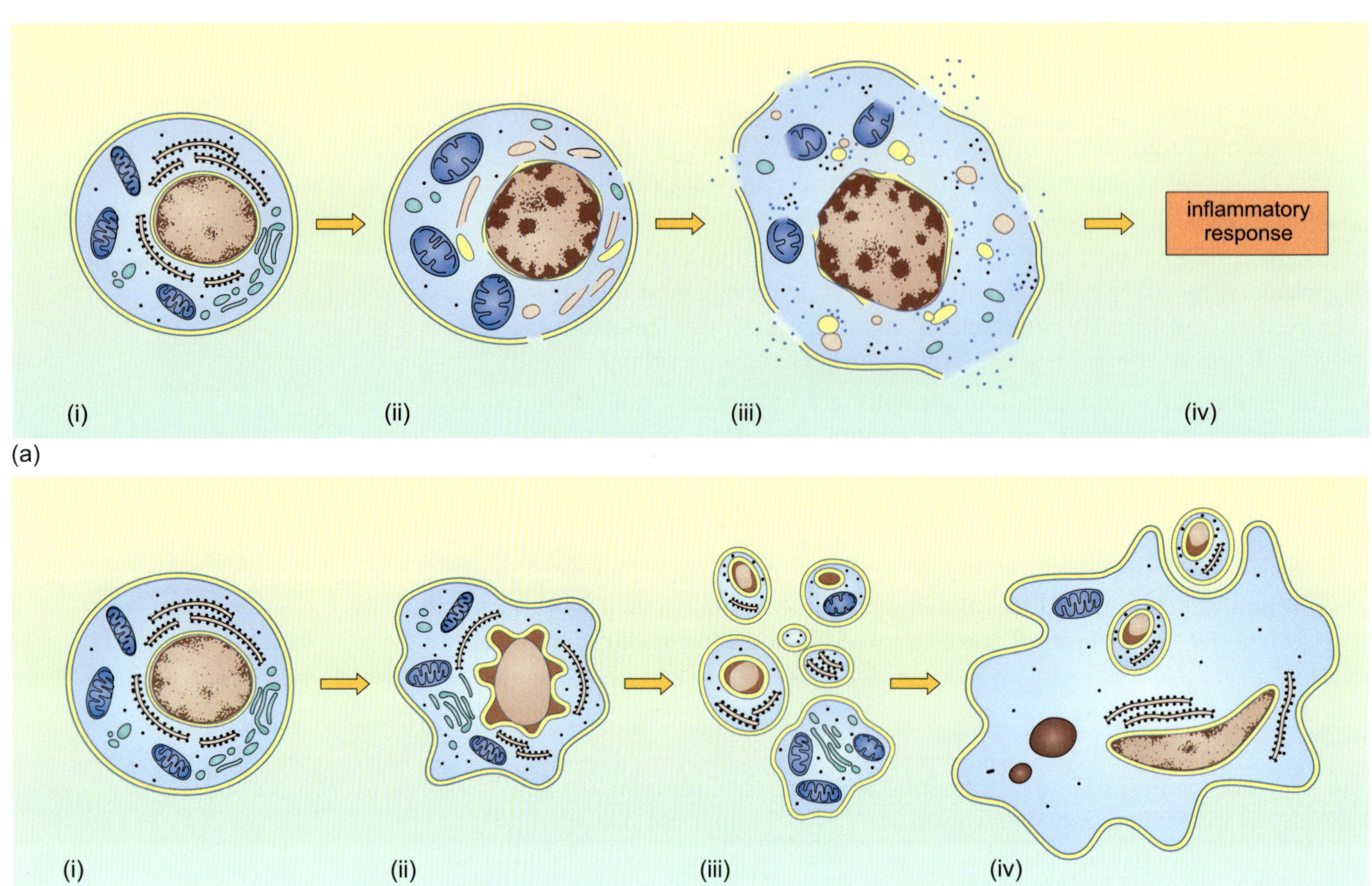

Figure 1.11 Schematic diagrams showing some of the events that occur in cells undergoing each of the two main forms of cell death. Note that these are events that can be detected by microscopy; molecular events are not shown. (a) Necrosis. (i) A normal cell. (ii) The first events of necrosis are the irregular condensation of the chromatin, swelling of the mitochondria and breakdown of membranes and ribosomes. (iii) The cell components are eventually completely disrupted and the cell lyses, releasing its contents; this event stimulates an inflammatory response (iv), during which macrophages infiltrate the tissue and engulf the remains of the cell (not shown). (b) Programmed cell death. (i) A normal cell. (ii) The first events that can be seen during programmed cell death are the involution of the nucleus, distinct condensation of the chromatin into large aggregates, and shrinkage of the cell. The other organelles appear normal at this stage. (iii) The nucleus and the cytoplasm break up into membrane-bound fragments known as apoptotic bodies, which are phagocytosed by nearby cells (iv).

The other type of death was first identified in certain stages of the development of organisms, where it was observed that particular cell populations die in a reproducible manner in every individual. Because of its predictable nature, this form of death was believed to occur as the result of a death 'programme', and so was named **programmed cell death**, also often called **apoptosis**. Well-known examples are the loss of the cells between the digits (e.g. during the development of fingers), and in the tail of the tadpole, when it metamorphoses into a frog. Apoptosis does not only occur during development, however. In adult tissues, cell death usually balances cell division, ensuring that tissues and organs retain the same size and structure as old cells are replaced.

The appearance of cells that are dying by apoptosis is very different from that of cells dying by necrosis. A cell dying by apoptosis shrinks and condenses (Figure 1.11b (ii)). The nuclear DNA breaks into fragments and forms aggregates, then the nucleus and cytoskeleton break up, resulting in the formation of a number of membrane-bound cell fragments called apoptotic bodies (Figure 1.11b (iii)). The apoptotic bodies are quickly engulfed by nearby phagocytic cells (Figure 1.11b (iv)). This process means the fragmented cell is very rapidly cleared away, and there is no inflammation because the cell does not lyse and release its contents. Thus, apoptosis is an active process that involves a particular sequence of intracellular events.

1.3.2   The apoptotic machinery

The mechanism of apoptosis is similar in all animal cells and involves a family of proteases called **caspases**, so named because they are **c**ysteine **asp**artases, cleaving proteins between adjacent cysteine and aspartate amino acid residues.

The caspases can be thought of as the cell executioners. Caspases of different sorts are present in cells all the time in an inactive form called a **procaspase**. In order to become an active enzyme, a procaspase requires proteolytic cleavage to reveal the enzyme's active site (Book 2, Section 1.6.2). There are several types of caspase in human cells, some of which are initiator caspases that cleave other caspases, and some are executioner caspases that cleave other cell protein components and trigger apoptosis. Once the first initiator caspase molecules are activated they cleave other caspases, thereby activating them and initiating a cascade of caspase activation (Figure 1.12). The activated 'executioner' caspases rapidly break down cell components. One of these caspases, for example, cleaves lamin proteins, causing breakdown of the nuclear lamina.

Clearly this proteolytic cascade is extremely destructive and must be kept under tight control. So why is the cascade triggered only in cells destined for destruction? In fact, several different pathways can converge to activate or inhibit caspases in different organisms.

The intrinsic pathway of apoptosis

One of the pathways that regulate caspase activity is known as the *intrinsic pathway*, because it is generated by signals internal to the cell, generally triggered by the detection of some sort of cell damage (Figure 1.13a). The intrinsic pathway involves members of the Bcl-2 family of proteins. When a cell is damaged, for example by DNA damage, virus infection or toxins, a Bcl-2 family member called Bax inserts into the outer membrane of the mitochondria causing them to release cytochrome *c* (which you will remember from Book 2, Section 3.6.3 is part of the electron transport chain) into the cytosol (Figure 1.13a). There, cytochrome *c* binds to molecules of a protein called Apaf-1 which aggregate together to form a complex called an

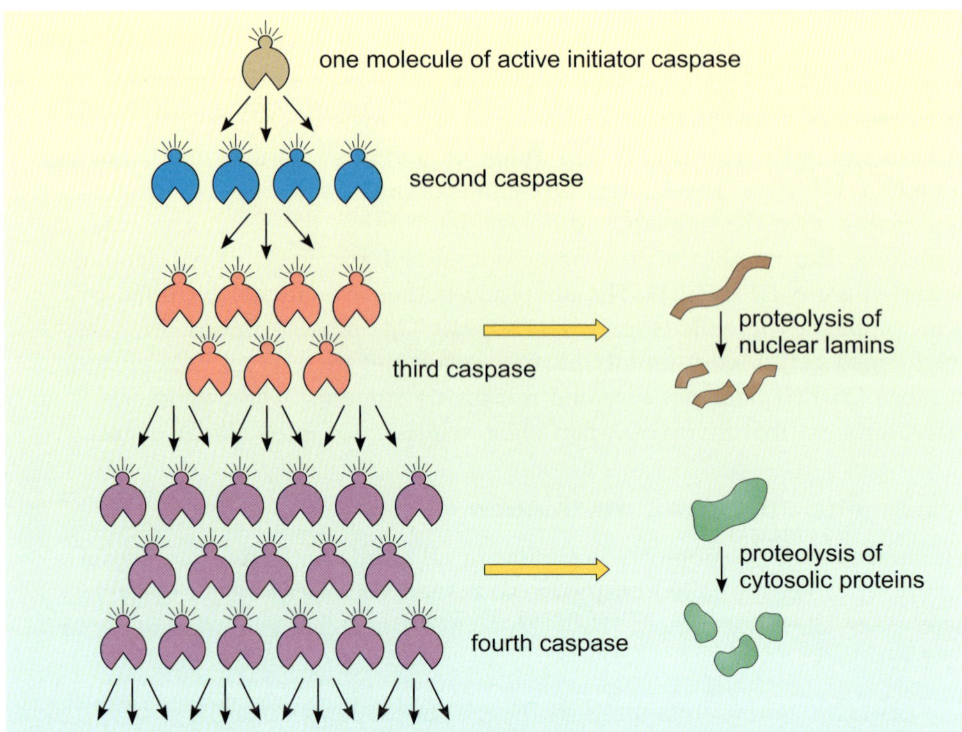

Figure 1.12 Schematic diagram showing the cascade of caspase activity that occurs during apoptosis. A cascade of proteolytic protein activity is set up in the cell following activation of only a small number of molecules of the first (initiator) caspase of the cascade (one molecule is shown). When activated, the initiator caspase cleaves many molecules of the inactive precursors of the second type of caspase in the cascade. The activated second caspase activates the third type of caspase in a similar manner; the activated third caspase breaks down nuclear lamins, as well as activating the fourth type of caspase. This final caspase causes proteolysis of cytosolic proteins.

apoptosome that cleaves and activates an initiator caspase, caspase 9, and thus triggers the caspase activation cascade.

In contrast, other members of the Bcl-2 family present in cells, including Bcl-2 protein itself, are anti-apoptotic, and they inhibit procaspase activation in various ways, such as by preventing the release of cytochrome c (Figure 1.13a). It is the balance between the activities of pro-apoptotic and anti-apoptotic Bcl-2 family proteins that largely determines whether a mammalian cell continues to live, or dies by apoptosis.

The process is, in fact, yet more complex. Another internal trigger is a molecule that you have already met in this chapter – p53.

- What is the role of p53 in the cell cycle?

- p53 is a transcription factor that is activated by damage to DNA. It halts progression through the cell cycle by, among other things, promoting the transcription of the cyclin–Cdk inhibitor p21 (Section 1.2.5).

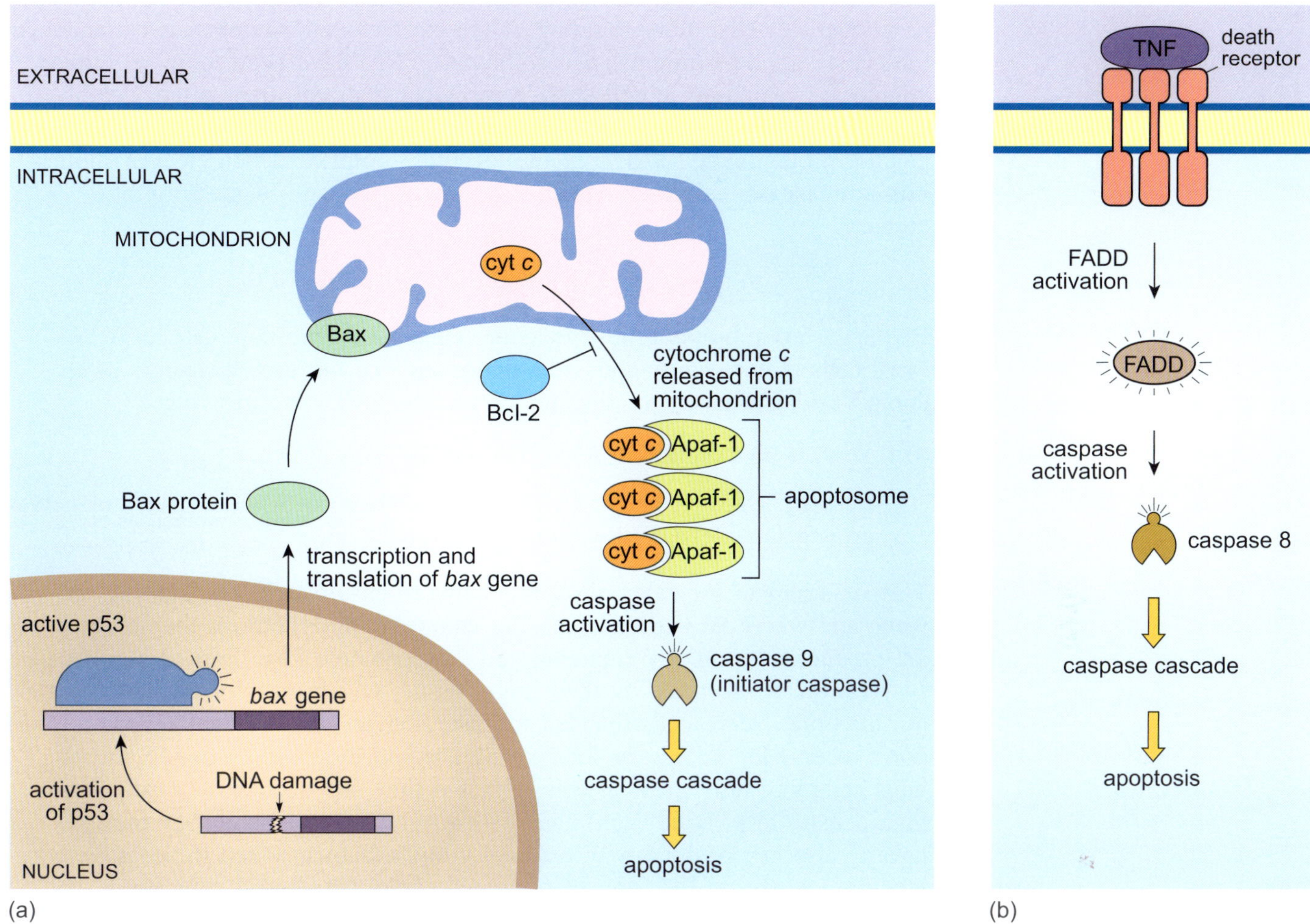

Figure 1.13 Intrinsic and extrinsic pathways leading to apoptosis. Several different pathways converge to activate or inhibit the caspases that are the mediators of apoptotic cell death. Only parts of these pathways are shown here. (a) The intrinsic pathway. In mammals, the pro-apoptotic protein Bax causes release of cytochrome c from the mitochondria, which activates the apoptosome and triggers the caspase cascade, resulting in apoptosis. This figure shows an example of severe DNA damage activating p53 protein, which in turn activates the transcription of pro-apoptotic proteins including Bax. In undamaged cells, anti-apoptotic proteins such as Bcl-2 inhibit activation of the apoptosome by, among other things, preventing cytochrome c release from the mitochondria. (b) The extrinsic pathway. Shown here is an example of the pathway, in which apoptosis is actively stimulated by extracellular signalling molecules such as tumour necrosis factor (TNF). Such factors act on so-called 'death receptors', activating the adaptor protein FADD, which results in activation of yet another initiator caspase, caspase 8.

This delay allows for DNA repair, but if the damage is too extensive the cell will eventually undergo apoptosis. In these circumstances, p53 activates the gene encoding the pro-apoptotic protein Bax, and thus promotes apoptosis via the intrinsic pathway (Figure 1.13a). Other molecules of the cell cycle can also play a role in determining the ultimate fate of a damaged cell, but there is not space to describe them all here.

The extrinsic pathway of apoptosis

Apoptosis can also be triggered by an external or *extrinsic pathway* that responds to signals from outside the cell, either through contacts with neighbouring cells, or via binding of growth factors. A number of extracellular molecules, such as tumour necrosis factor (TNF) have been found to induce apoptosis by binding to cell surface receptors that are often called 'death receptors' (Figure 1.13b). Binding of TNF to a death receptor recruits an adaptor protein called FADD to the receptor complex, which cleaves another initiator caspase, called caspase 8, leading to induction of the apoptotic cascade.

Prevention of cell suicide by survival signals

Balancing the pro-apoptotic signals are signals that promote cell survival. Most cells require continuous stimulation either from growth factors or by contact with other cells or the surface on which they are growing.

- ■ What is an example of a survival factor?

- ☐ Epidermal growth factor (EGF), insulin-like growth factor (IGF) or nerve growth factor (NGF) (Section 1.2.8, Table 1.1).

This requirement for external signals helps to ensure that cells survive only when and where they are needed. For example, more neurons than are needed are produced during development, and they compete for a limited number of survival factors secreted by their target tissue, which may for example be a muscle fibre. Nerve cells that fail to make contact with a target cell therefore don't receive the necessary survival signals and die by apoptosis. Similarly, most cells will only grow in cell culture if they are 'seeded' at a sufficiently high density to provide each other with survival signals, and they must usually also be provided with external growth factors, added to the growth medium, usually in the form of separated serum (the liquid component of blood from which cells have been removed). Survival factors bind to cell surface receptors and turn on signalling pathways that suppress apoptosis, usually by regulating the intracellular concentration of Bcl-2 family proteins. Whether a cell lives or dies thus depends upon the interplay between many different molecules.

Summary of Section 1.3

- Most cells die by one of two distinct mechanisms: necrosis or apoptosis (programmed cell death).
- Necrosis is the response of a cell to mechanical damage or toxic substances and involves cell lysis, release of cell contents and inflammation in mammals.
- Apoptosis is essential both during development and in normal tissue maintenance. It involves destruction of DNA and structural components of the cell; the cell does not release its contents and there is no inflammatory response. The apoptotic cell is phagocytosed by other cells.
- Caspases are the enzymes that, when activated, carry out the destruction of the cell. Caspases exist in cells as inactive precursors, but when activated cleave proteins including other caspases, initiating a cascade of proteolytic activity.

- The death programme can be activated or inhibited by a variety of intra- and intercellular stimuli that together determine the fate of a cell.

1.4 Cell differentiation

The bodies of higher vertebrates, including humans, contain more than 200 different types of cells which all develop from a single-celled zygote (Book 1, Chapter 4). The process by which non-specialised cells, like those in the early zygotic cell divisions, develop into specialised cell types is differentiation. What are the processes that determine what cell type a cell will ultimately become?

Much of the current understanding of cell differentiation has come from the study of *embryogenesis*, the formation and development of the embryo in animals and plants. It was thought for many centuries that embryogenesis consisted entirely of growth as a result of cell proliferation; but we now know that the development of organisms is extremely complex, involving overlapping phases of cell growth, division, migration, differentiation and programmed cell death, which culminate in morphogenesis – the formation of the organism's characteristic shape and specialised structures and tissues.

■   What underlies the differences between cell types in a multicellular organism?

□   Differential gene expression, which results in the synthesis of a different set of proteins in different cell types (Book 1, Section 2.4.5).

Each specialised cell type in a multicellular organism expresses a subset of the genes encoded by the genome. During differentiation, a cell changes its pattern of gene expression and may also undergo dramatic changes in morphology, metabolic activity, and responsiveness to signals.

■   What are the molecules that most directly regulate gene expression called?

□   Transcription factors.

In eukaryotes, the transcription of a gene is under the control of a number of different types of transcription factor; thus gene expression is said to be under *combinatorial control* (Figure 1.14 and Book 1, Section 6.4.3). In addition, each type of transcription factor can participate in regulating many different genes, so it is possible to generate a wide range of biological complexity using different combinations of relatively few regulatory proteins. During differentiation, a cell might sequentially switch on expression of a number of transcription factors, but it may not experience large changes in the pattern of gene expression until a final key regulator is expressed.

Over the last few decades, research has revealed that despite the great diversity of animals and plants, a very similar group of genes, proteins, mRNAs and signalling pathways drive their early development. The details of these mechanisms are complex, so a full description of the development of even a simple organism is beyond the scope of this text. This section will instead outline the general cellular and molecular mechanisms that result in

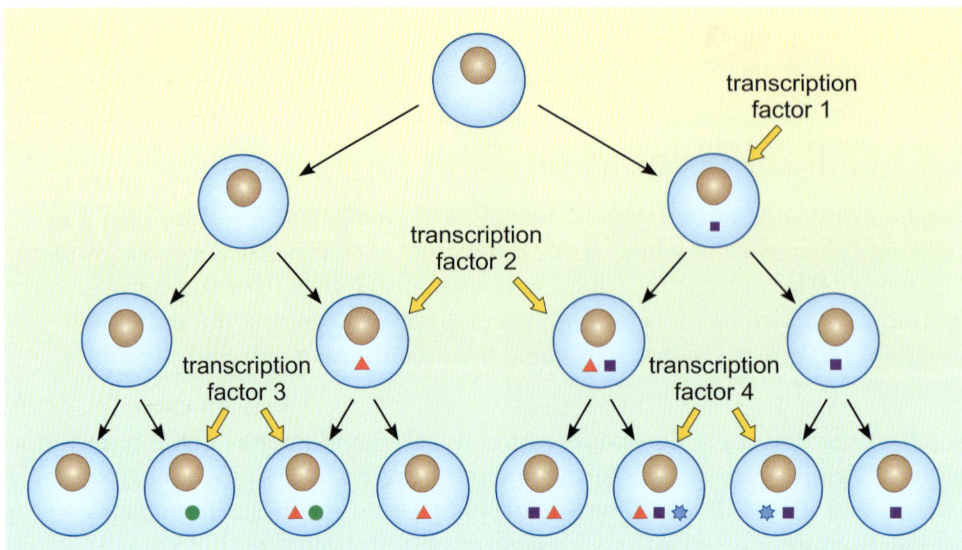

Figure 1.14 Combinatorial control of eukaryotic gene expression. Gene expression depends upon the presence of several different transcription factors. Exposure of a cell to different combinations of transcription factors thus results in expression of different genes and a different complement of proteins, shown as different-coloured shapes within the cells. Differential expression of genes, under combinatorial control, results in the development of different cell types.

cell differentiation and the determination of **cell fate**, focusing on examples from animal development. Remember also that cell differentiation does not occur only during development. Some types of cells, including those of the immune system and the epithelia of the skin and intestine, undergo proliferation and differentiation throughout life in order to maintain tissue integrity and function. Differentiation in mature organisms is considered briefly in Section 1.4.4.

1.4.1 Development in multicellular eukaryotes

In animals, development begins when an egg is fertilised by fusion with a sperm cell to form the single-celled zygote which is **totipotent**; that is, it has the potential to form an entire organism. Development has been extensively studied in several animal model organisms including the mouse (*Mus musculus*), the African clawed frog (*Xenopus laevis*) and the fruit fly (*Drosophila melanogaster*) (Figure 1.15).

These animals have very different types of eggs, and there are many differences in the arrangement and movement of cells during development. Nevertheless, it is possible to distinguish similarities in their early development. In most animals, the zygote initially undergoes a series of mitotic divisions known as cleavages (because no cell growth occurs between them) that give rise to a **blastula** (or a blastocyst in mammals), which is a hollow sphere of cells surrounding a fluid-filled cavity. A group of cells (called the inner cell mass) inside the blastula or blastocyst will ultimately form the embryo and they are **pluripotent**; they are no longer individually able to give rise to a whole organism, but can still form virtually any type of

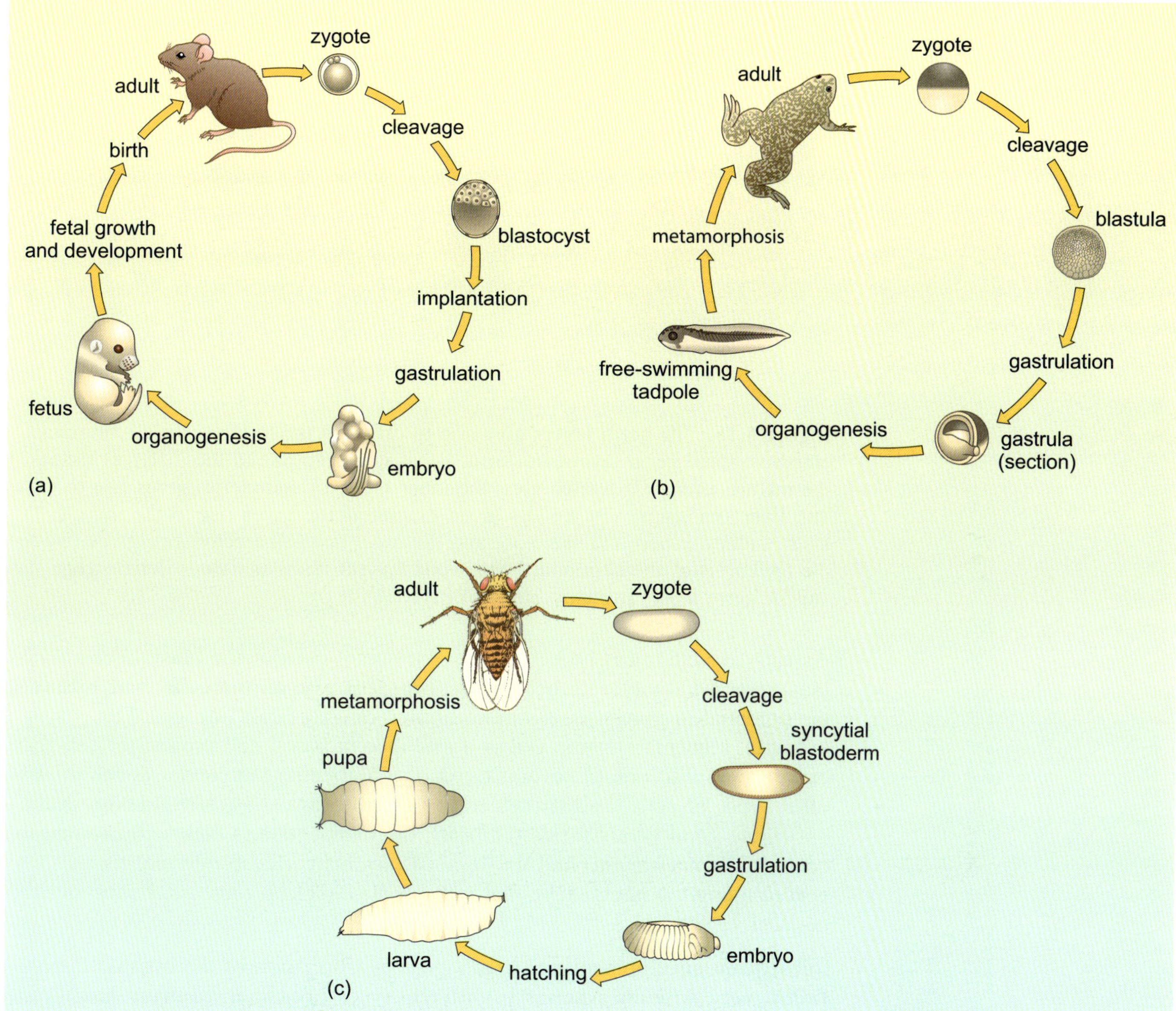

Figure 1.15 Simplified view of the main stages of development of three model organisms, not to scale. (a) Mouse, *Mus musculus*; (b) frog, *Xenopus laevis*; (c) fruit-fly, *Drosophila melanogaster*. Note that the development of these organisms takes varying times. Zygotes vary greatly in size and shape, as indicated here, but usually go through a period of rapid cell division which results in the formation of a 'ball' of cells known as the blastula or blastocyst, except in the case of *Drosophilia*, where rapid nuclear division without cytoplasmic division occurs, resulting in the formation of a syncytial blastoderm (as described in Box 1.2). Different layers of cells develop from the cells of the blastula or blastocyst during the cell movements and interactions known as gastrulation (see text).

cell in the body. This property, as you will see later, has led to their use in the development of stem cell technology.

The blastula or blastocyst then begins to undergo **gastrulation** (forming the *gastrula*) which is characterised by the invagination (inward movement) of the outer cells into the centre, resulting in the formation of three **germ layers**, which will ultimately give rise to the different tissues of the mature animal

(Figure 1.16). By this stage, the basic 'pattern' of the body (i.e. the anterior and posterior (head and tail), dorsal and ventral (back and front) and left to right axes of the organism) is already beginning to be determined.

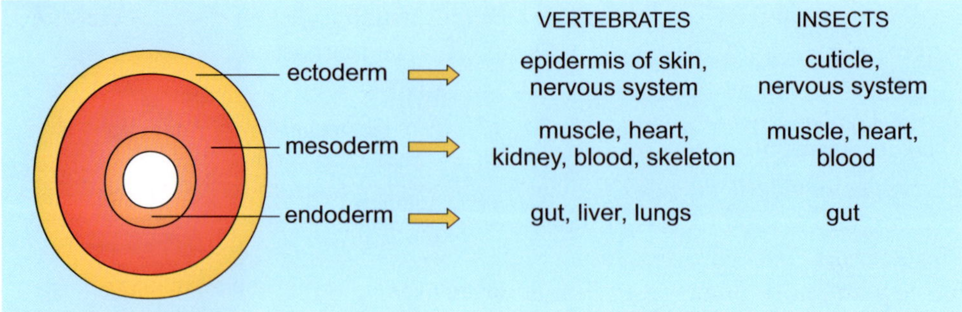

Figure 1.16 The germ layers give rise to different tissue types. In all animal embryos, three germ layers are formed that will give rise to the different tissues of the mature animal. In vertebrates, the inner layer, or endoderm, gives rise to the gut, liver and lungs. The middle layer, or mesoderm, gives rise to the muscle, heart, kidney, blood and skeleton, and the outer layer, the ectoderm, gives rise to the epidermis of the skin and the nervous system. Similar tissues derive from similar germ layers in many kinds of invertebrates.

As early development proceeds, the pluripotent cells of the blastula or blastocyst differentiate to become **multipotent progenitor cells**, which have the potential to give rise to a more limited range of cell lineages. For example, a haematopoietic cell is a multipotent blood cell progenitor cell that can develop into one of several types of blood cells, but cannot develop into other types of cells. Further cell migration, division and differentiation in the vertebrate germ layers forms the heart, the neural tube (which will become the central nervous system) and the somites (segments of mesodermal tissue spaced along the neural tube that will become skeletal muscles, bones and connective tissue).

Subsequent development and organogenesis (formation of the organs) then follow very different patterns in different types of animal. In many insects and amphibians, for example, embryonic development results in an intermediate larval form, which must undergo metamorphosis into the adult form (there is no corresponding larval stage or metamorphosis in mammals, Figure 1.15). Many complex interactions between cells, cell movements and even cell death are required to achieve the final spatial organisation of different tissues and organs.

Plants also develop from a zygote by cell division and differentiation to generate different cell types. However, while the morphology of an animal is determined by the migration of cells during gastrulation to form the three germ layers (ectoderm, mesoderm and endoderm), plant cells are unable to migrate in this way.

■   Why are plant cells unable to migrate?

☐   Plant cells are trapped in rigid cellulose walls, which prevent their movement; that is, they are positionally fixed after division.

So, while plants also form three major types of tissue (dermal, ground and vascular), their overall form depends only on the rate of division of the different types of cell, and the direction of elongation of tissues. Also, while the basic body plan of animals is 'fixed' by gene expression early on in development, and external influences cannot usually change this predetermined pattern, plants don't have a predetermined body plan; instead their development is continuous. New plant tissues are formed throughout the plant's life by clusters of undifferentiated cells called **meristems**, which are found in particular zones where growth takes place. Plants therefore have much greater developmental plasticity (the ability to alter their development in response to external factors, such as light, temperature or nutrient availability) than animals. The meristem can also regenerate damaged parts (or in some species, an entire plant) from individual cells.

1.4.2   The regulation of differentiation during development

Cell diversity in all multicellular organisms is achieved in two main ways. Firstly, a cell may undergo *asymmetric cell division*, to produce two daughter cells that inherit unequal amounts of intrinsic cytoplasmic molecules, known as **cell fate determinants**, from the parent cell. Secondly, two initially identical cells may become different if they are exposed to different extracellular signals from surrounding cells.

Asymmetric cell division

What you have learned about cell division so far has assumed that it always gives rise to two identical daughter cells, but many cell divisions are in fact asymmetric. **Asymmetric cell division** generates two daughter cells that differ in their ability to develop into particular cell types.

The cytoplasm of an unfertilised oocyte (immature egg cell) is already packed with molecules synthesised by the mother, including large amounts of transcription factor proteins, and mRNAs encoding transcription factors. When the egg is fertilised, these maternal transcription factors will direct gene expression in the early stages of development, before the zygotic DNA genome starts to be expressed. While an oocyte may appear homogeneous, at the molecular level it is in fact highly asymmetrical, with an uneven distribution of maternally deposited cell fate determinants including transcription factors and other types of regulatory molecules.

After fertilisation, the zygote initially undergoes a series of cleavages (Figure 1.15). The daughter cells derived from one pole of the oocyte may thus receive most or all of certain cytoplasmic cell fate determinants, while the daughter cells derived from the other pole receive less or none. This therefore results in daughter cells which start off with different amounts of transcription factors and consequently have different cell fates based on differences in gene expression. In the nematode worm *Caenorhabditis elegans*, the very first cleavage of the zygote produces an anterior and a posterior cell containing different amounts of the cell fate determinants called PAR proteins. The unevenly distributed PAR proteins are members of signalling pathways that orchestrate a whole series of asymmetric cleavages that determine the anterior–posterior, dorsal–ventral and left–right axes early in the development of the *C. elegans* body plan.

Perhaps the best studied example of the role of unequal distribution of cell fate determinants in the establishment of the body plan is the early development of the model organism *Drosophila* (briefly explained in Box 1.2).

Asymmetric cell division also has a role in later stages of development and has been extensively studied in the development of muscle, gut and the nervous system. For example, *Drosophila* neurons arise from progenitor cells called *neuroblasts* which divide asymmetrically to give rise to two cells with different fates: another neuroblast and a ganglion mother cell (GMC) (Figure 1.17a). This ability to 'self-renew' by dividing (by mitosis) to give one new progenitor cell and one new specialised cell is a characteristic of stem cells, which will be discussed in Section 1.4.4. The new neuroblast will carry on going through repeated divisions, but the GMC divides just once more to produce two fully differentiated neurons. The cell fate determinants in the dividing *Drosophila* neuroblast include a protein called Prospero, a transcription factor that activates neuron-specific genes and represses neuroblast-specific genes (the *prospero* gene is named after the wizard who controls the other characters in Shakespeare's play, *The Tempest*).

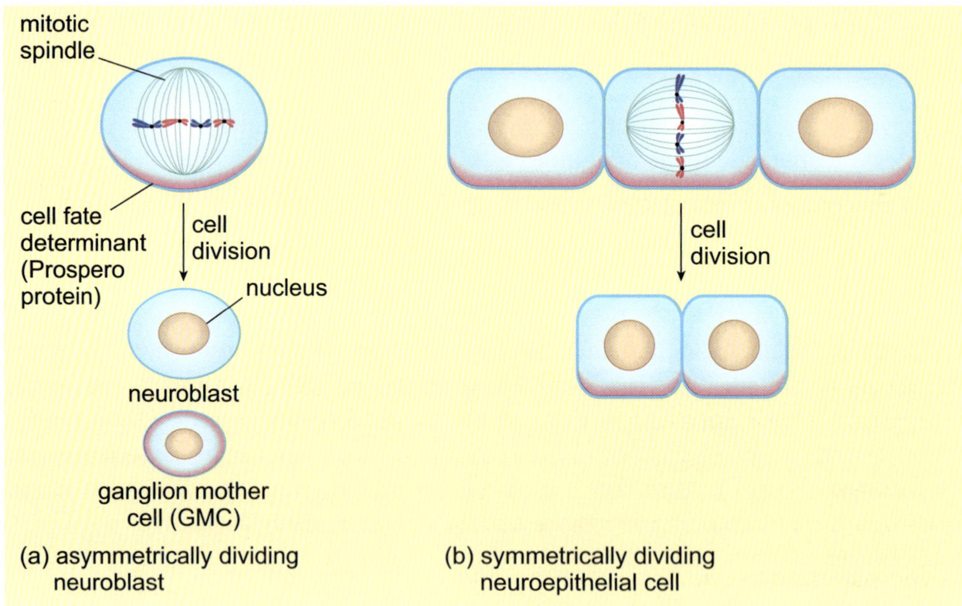

Figure 1.17 Diagram showing how asymmetric cell division generates cell diversity. (a) Asymmetric cell division in *Drosophila* neuroblasts results in uneven distribution of cytoplasmic cell fate determinants which drive the daughter cells to go on to follow different developmental pathways. The cell that inherits the Prospero protein becomes a ganglion mother cell, while the other cell becomes a new neuroblast. (b) Neuroepithelial cells, in contrast, undergo equal partitioning of Prospero to produce two new neuroepithelial cells. The spindle orientation is crucial for the outcome of division.

Another protein called Miranda (named after Prospero's daughter in *The Tempest*) tethers the Prospero protein to the basal side of the neuroblast, which, when asymmetric division occurs, will become the GMC, while the apical side becomes the new neuroblast (Figure 1.17a). In order for cellular determinants like Prospero to be segregated correctly into only one of the daughter cells, there is therefore a second requirement: that the cell must also orient the mitotic spindle in such a way that cleavage occurs in a particular orientation.

■ During cytokinesis, where does cell cleavage (cytokinesis) occur in relation to the spindle microtubules?

☐ Cleavage of the daughter cells is always perpendicular to the long axis of the mitotic spindle (Section 1.2.7).

In Figure 1.17a, the mitotic spindle orientation is perpendicular to the basal side of the cell (where the cell fate determinant protein, Prospero, is located) resulting in two daughter cells with an unequal distribution of Prospero. One daughter cell lacks Prospero and remains a neuroblast, while the other Prospero-containing daughter cell develops into a GMC. In contrast, neuroepithelial cells (a different type of progenitor cell found in the same tissues as neuroblasts) undergo symmetrical division. Here, the mitotic spindle is *parallel* to the basal side of the cell, with the result that the two daughter cells inherit an equal amount of Prospero protein (Figure 1.17b), so both will become identical neuroepithelial cells.

In theory, it should be possible to generate an entire organism by asymmetric cell division, the fate of each cell depending upon its lineage; but although lineage plays an important role in development, it cannot alone produce a differentiated organism, and determination of cell fate is far more often the result of signals received from other cells.

Box 1.2 Cell fate determinants in *Drosophila* development

Drosophila is a segmented animal; its body plan consists of repeated segments along the anterior to posterior (head to tail) axis. Each segment will develop different specialised structures such as legs or antennae. Segmented features are also present, although less obvious, in other animals: for example, the vertebrae and ribs in mammals.

The pattern of segments along the future anterior–posterior axis of *Drosophila* starts to be determined during egg formation, long before fertilisation. The cytoplasm of the developing egg is polarised by an uneven distribution of **maternal mRNAs** laid down in the egg by the mother. These include an mRNA transcribed from a gene called *bicoid*. There is concentration gradient of *bicoid* mRNA along the length of the egg; higher at the anterior (head) end and lower at the posterior (tail) end (Figure 1.18). When the egg is fertilised, the *bicoid* mRNA is translated, so a similar gradient of Bicoid protein is established. The Bicoid protein is a transcription factor required for transcription of the first zygotic genes to be expressed (which are known as *gap genes*). Importantly, gap genes are only transcribed where there is an appropriate level of Bicoid protein. The Bicoid protein gradient (and gradients of other maternally encoded proteins) thereby provides a positional cue that results in expression of the zygotic gap genes only in certain zones of the developing zygote.

Early embryogenesis in *Drosophila* differs from vertebrate animals in that the early cleavage stages don't result in a ball of individual cells. Instead, the nuclei divide, but continue to share a single large cytoplasm, a stage called the syncytial blastoderm (Figure 1.18). Essentially, the syncytial blastoderm is a single cell with over 8000 nuclei spaced around the single cell membrane. As a result, the gap proteins are translated in the shared cytoplasm and can freely diffuse to form concentration gradients that regulate gene expression in nearby nuclei. The products of gap genes are transcription factors and their uneven expression sets up further zones of localised gradients that activate nearby nuclei to transcribe the genes for another set of transcription factors, the *pair-rule genes*, which (as you can see from Figure 1.18) are expressed in seven repeated stripes along the blastoderm. The transcription factors expressed by the pair-rule genes in turn activate transcription of the *segment polarity genes* which are expressed in 14 repeated stripes. The gap genes, pair-rule genes, and segment polarity genes are collectively called **segmentation genes**, because they are involved in setting up the location of body segments.

The final level in this gene expression hierarchy is the activation of *homeotic genes*, also known as *Hox genes*. These genes encode transcription factors that regulate a variety of target genes that will determine the morphology and structure of each segment (i.e. whether it will become part of the head, thorax or abdomen of the fly). By this point, the blastoderm has begun to 'cellularise'; membranes have formed around the nuclei to give individual cells and gastrulation occurs to give the three germ layers (Figure 1.18). Hox genes were first identified in *Drosophila* mutants in which inactivated Hox genes resulted in dramatic defects in particular body segments. For example, mutations in a Hox gene called *antennapedia* cause legs to grow in the place of antennae on the insect's head. Hox genes are highly conserved and are present in all other animals. Intriguingly, Hox genes appear to be arranged in clusters in the genome in the order in which they are expressed along the body axis.

This example in *Drosophila* illustrates how an initial cytoplasmic gradient of a transcription factor, already present in the unfertilised egg, can switch on a complex sequence of 'master' regulatory genes (also encoding transcription factors) in different regions of the zygote. The overlapping patterns of transcription factor expression act locally to determine different fates for cells in different parts of the embryo and establish complex body plans.

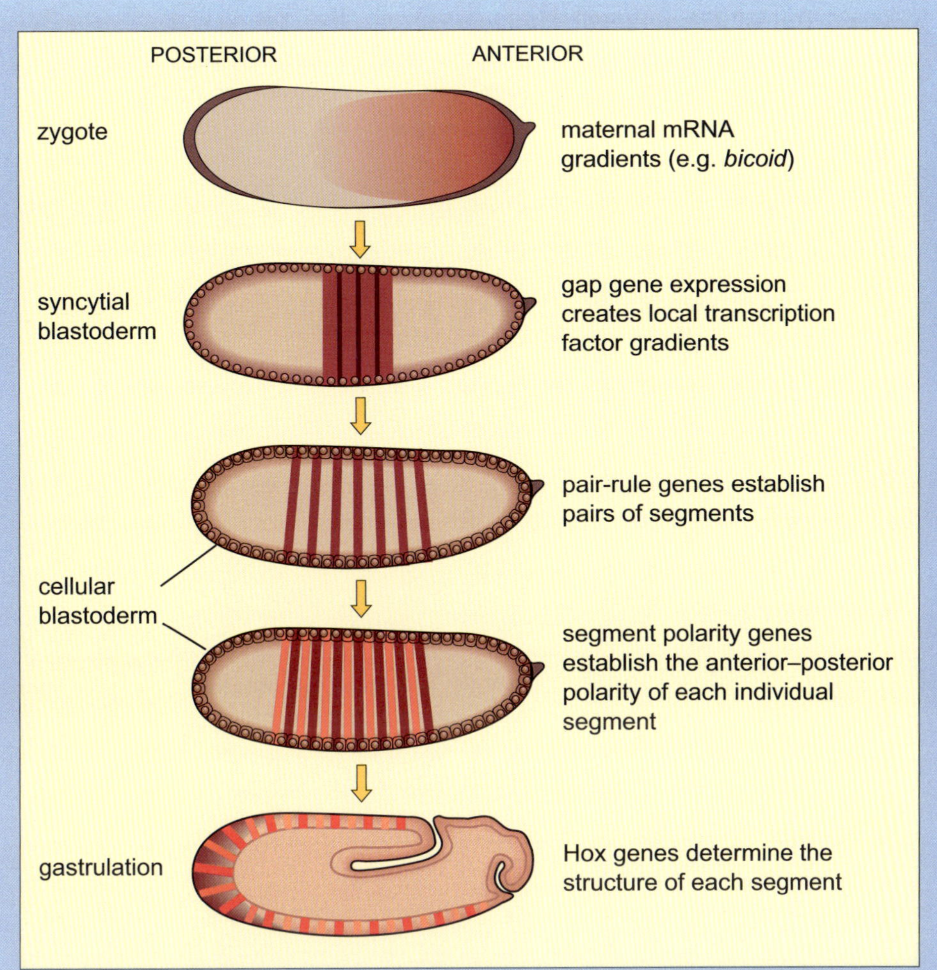

Figure 1.18 Gradients of maternal mRNAs in the *Drosophila* zygote initiate the localised expression of a hierarchy of regulatory genes that set up the pattern of anterior–posterior segments in the adult. The concentration of maternal *bicoid* mRNA is higher at the anterior end of the oocyte. Translation of *bicoid* mRNA after fertilisation leads to a gradient of the Bicoid transcription factor, which results in localised transcription of the zygotic gap genes. Gap genes also encode transcription factors, and localised gradients of these proteins results in the localised expression of pair-rule genes. The sequential activation of segment polarity genes followed by Hox genes ultimately results in segmentation of the body, and the development of different structures in each segment.

Signalling between cells

As cells migrate during development they are brought into contact with different groups of neighbouring cells. In fact, the great majority of molecular processes that control cellular differentiation probably involve signalling between cells.

■   What types of cell interactions can convey signals between cells?

□ Direct cell–cell or cell–substrate contact through cell surface receptors or cell adhesion molecules (e.g. integrins and cadherins); diffusible signals that are released from one cell and bind to receptors on another (e.g. growth factors); and diffusible signals that move from cell to cell via gap junctions.

You met examples of these in Book 2, Chapter 4, and in fact all of these types of signalling pathways play a role in the differentiation of cells.

■ What type of intercellular signalling involves diffusion of a secreted signalling molecule to nearby target cells?

□ Paracrine signalling (Book 2, Section 4.2.2).

In order for a signalling molecule to exert an effect on specific target cells rather than all the cells it might reach, there must be a means of generating different responses to the molecule in different cells.

■ Suggest four reasons why two cells might have a differential response to a signalling molecule.

□ Differential responses of cells to a signalling molecule could occur because of:

i   exposure to different levels of the molecule

ii   differential expression of receptors for the molecule

iii   differential expression of intracellular molecules needed to transduce the signal

iv   differential exposure to an inhibitory molecule (e.g. an antagonist that prevents the action of the signalling molecule in some way).

An example of intracellular signalling is in the development of germ layers during the initial stages of vertebrate development (Figure 1.16). Cells in the two poles of an amphibian blastula, known as the vegetal pole and the animal pole, inherit differing amounts of cell determinants (in this case, maternal mRNAs) in the early divisions and go on to form different germ layers in the embryo. The vegetal pole cells form the endoderm (which eventually forms the gut and some other internal organs) while animal pole cells form ectoderm (which forms the skin and nervous system). However, cells of a third type form between these two germ layers, and develop into mesoderm, which gives rise to muscle, bone and other tissues. Painstaking experiments were performed in which very small pieces of tissue from the animal and vegetal poles of the late blastula stage of *Xenopus laevis* were grown as explants in culture, either together or separately. These experiments showed that when the animal pole cells and vegetal pole cells were grown separately, mesoderm did not develop. Interactions between the two cell types are thus needed for mesoderm to be formed (Figure 1.19). It turns out that the mesoderm is formed from animal pole cells, as a result of signals they receive from vegetal pole cells. Such interaction, in which the developmental fate of one cell type is determined by signals from adjacent cells, is known as **induction**.

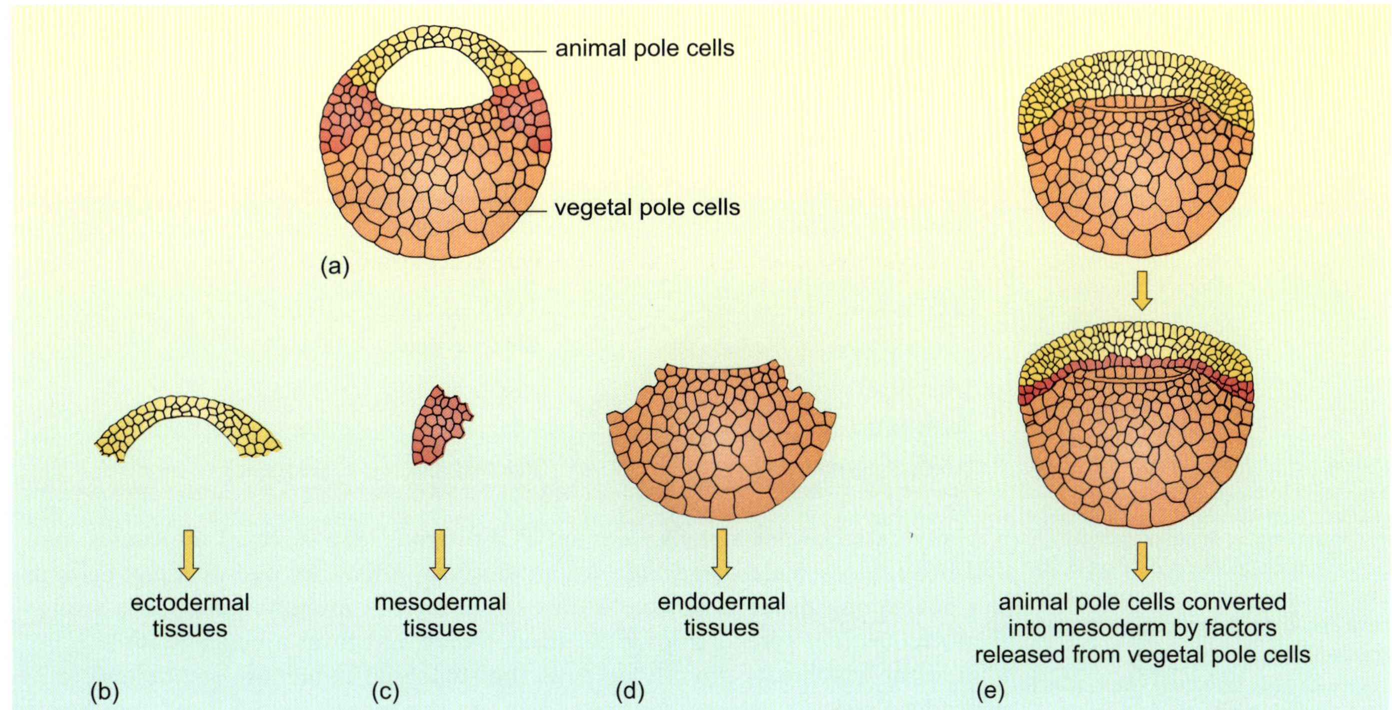

Figure 1.19 Induction of mesoderm in the frog *Xenopus laevis*. (a) In a normal embryo, mesoderm forms from cells that lie between animal and vegetal pole cells. When cells from different regions of a late blastula are cultured alone, they follow different developmental paths: (b) animal pole cells form only ectodermal tissues; (c) cells from the region between animal and vegetal poles develop into mesodermal tissues; (d) vegetal pole cells only give rise to endodermal tissues. (e) When animal and vegetal pole cells from an early blastula are cultured next to each other, mesodermal tissue develops from the animal pole cells.

In further experiments, the explanted animal pole tissues were bathed in tissue culture medium in which cultured *Xenopus* cells had previously been grown. This treatment induced the formation of mesoderm, showing that the factors responsible were secreted molecules that reached their target cells by diffusion. The secreted factors were subsequently purified from the culture medium and identified. Evidence suggests that several molecules are in fact involved. One is a polypeptide growth factor in the transforming growth factor β (TGFβ) family (Table 1.1) which can trigger differentiation of all the cells in a particular region by activating a signalling pathway that alters gene expression.

In fact, induction in animal development is controlled by a relatively small number of conserved signalling pathways, which are employed over and over again in the development of many different structures and organs, including limbs, teeth, facial structures, hair follicles, kidneys, lungs, and gut. Some of the major signalling pathways are listed in Table 1.2. The ultimate fate of a cell depends upon the integration of several signalling pathways, as well as the cell's 'memory' of signals that it has previously received, all of which combine to induce the transcription of a particular set of genes. A limb therefore ends up looking very different from a tooth, even though the molecular mechanisms that have specified them may be similar.

Table 1.2 Examples of signalling pathways that induce differentiation in many animal tissues. Most pathways involve families of related receptors and ligands; only some examples are shown here.

Pathway	Receptors	Family of extracellular signalling molecules
Receptor tyrosine kinase pathway	EGF receptor	EGF
	FGF receptor	FGF
TGFβ pathway	TGFβ receptor	TGFβ
	BMP receptor	BMP
Notch pathway	Notch	Delta
Wnt pathway	Frizzled	Wingless (Wnt)
Hedgehog pathway	Patched	Hedgehog

In fly embryos, transcription factors such as Bicoid (Box 1.2) can be considered morphogens because they form concentration gradients that act directly on nuclei in the syncytial blastoderm to determine development of tissues in the embryo.

Many extracellular signals, like the TGFβ family, are secreted molecules that diffuse into a widespread area where they can affect the fate of many cells in a developing tissue. Such molecules are known as **morphogens**; these are molecules that spread from a localised source and form a concentration gradient which acts *directly* on cells (not by serial induction, see below) to produce specific cellular responses that are dependent on the local morphogen concentration.

A well-studied example of this is the Hedgehog family of extracellular signalling factors (Table 1.2). One member, Sonic Hedgehog (Shh) is required for anterior–posterior patterning during animal limb development, which begins when cells originating from the mesoderm germ layer (Section 1.4.1) start to proliferate and bulge outwards forming a limb bud. As the limb bud elongates, a small region of cells called the zone of polarising activity (ZPA) on the posterior side of the developing limb bud secretes Shh, which diffuses away from the cells that make it and stimulates mesodermal cells throughout the limb bud to form the digits. The Shh gradient thereby provides positional information to cells. Cells close to the ZPA are exposed to a high concentration of Shh and form posterior digits, while cells further away are exposed to a lower concentration and form anterior digits (Figure 1.20a). If a piece of ZPA tissue, or a bead coated in Shh, is implanted in the anterior side of a chick limb bud, the limb will develop a mirror-image duplication of the digits (Figure 1.20b). The final moulding of digits in the paddle-shaped limb bud is achieved by another secreted factor, BMP (Table 1.2), which triggers apoptosis of the cells between the digits.

Overlapping gradients of morphogens along the three different axes of the body (anterior–posterior, dorsal–ventral and left–right) are thought to provide three-dimensional positional information which informs cells exactly where they are located and how they should develop.

While a morphogen gradient, whether it be mRNAs as in *Drosophila* segmentation (Box 1.2) or extracellular signalling molecules such as Shh, can set up the initial basic pattern of the organism, cells require other information in order to determine the fine structure of tissues, for example the polarity of an individual epithelial cell within a gut epithelium (Book 1, Section 2.4.5).

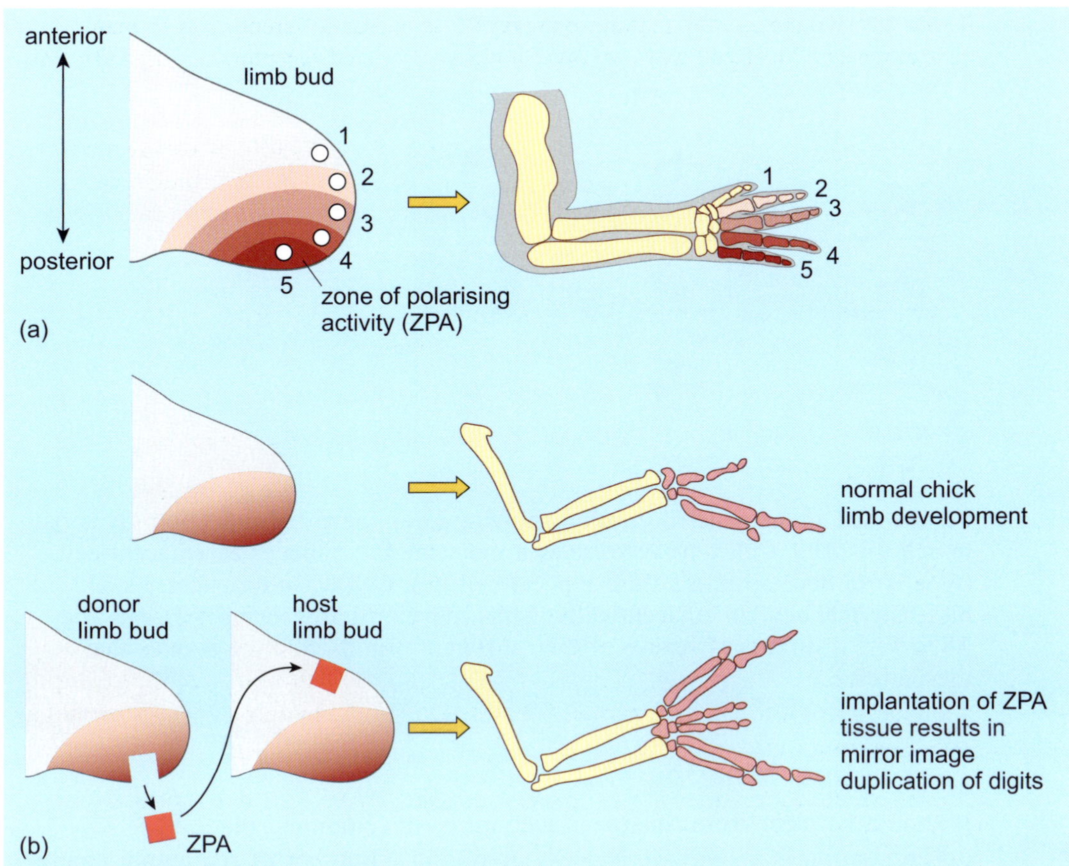

Figure 1.20 Morphogen gradients can determine the development of limb structures. (a) Morphogens, including Shh, are secreted from the posterior zone of polarising activity (ZPA) and diffuse through the limb bud. Exposure to the concentration gradient of morphogens (indicated by the intensity of colour) determines the type of digits that will form at each position. (b) Anterior implantation of ZPA tissue, or a bead coated in Shh, into the developing chick limb bud results in mirror-image duplication of the digits.

This level of tissue architecture is probably determined locally by interactions between cells that lie next to each other after division, or that are brought together by cell movements during development. This so-called **sequential induction** arises from a series of signals between cells, where a signal from one cell determines the fate of a neighbouring cell, which in turn induces the differentiation of a third type of cell, and so on (Figure 1.21).

Some inductive signals are very short-range, notably those transmitted via cell–cell contacts: for example, the Notch signalling pathway (Table 1.2) which is involved in the differentiation and development of many animal tissues, particularly those of the nervous system. Notch is a transmembrane receptor protein that interacts with a transmembrane ligand called Delta on an adjacent cell. When the receptor and ligand come into contact, the Notch receptor is cleaved and the intracellular domain thus released moves to the nucleus to control gene expression (Book 2, Figure 4.3). This signals to the Notch-bearing precursor cell not to differentiate into a nerve cell. In the vertebrate eye, the interaction between Notch and its ligand regulates which cells become optic neurons.

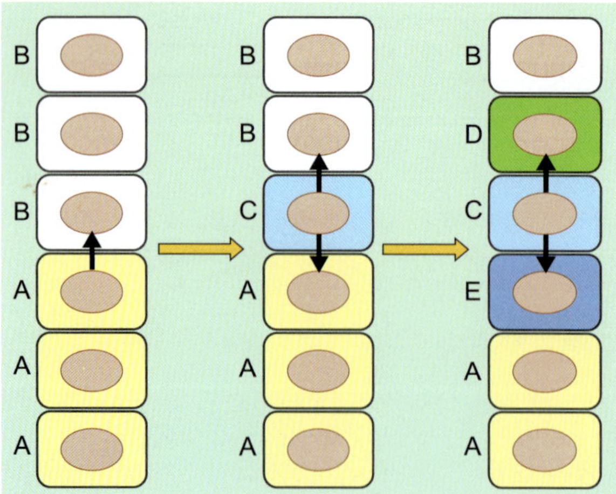

Figure 1.21 Schematic illustration showing how sequential induction gives rise to cell diversity. Cell type A secretes a molecule that induces an adjacent cell (type B) to develop into a third type of cell (C). Cell C, in turn secretes a molecule that acts on adjacent cells, which develop into cell types D and E. Their fate is different because of their different origins, and the signals that they receive.

1.4.3 Determination and flexibility

It should be clear from this brief account of development that cell differentiation is a stepwise process involving a number of sequential changes in cells as they assume their fully differentiated state. It is important to appreciate this temporal aspect of differentiation. At any point in time, the nature of a cell and its possible future characteristics depend upon its developmental history. While a zygote is capable of giving rise to all types of cell, as an embryo develops the developmental potential of the cells gradually becomes more restricted. This process can be divided into three states. Most cells first become **committed** to follow a particular developmental pathway, but a cell's fate can be reversed or changed at this stage. At the next stage the progenitor cell becomes **determined** to differentiate into particular cell type. Its fate cannot be changed at this stage. For example, a cell that has become determined to become a muscle cell cannot become a skin cell. Determination is followed by the actual changes in cell biochemistry, morphology and function that result in differentiation into a specific cell type.

An example of this process is the development of skeletal muscle in vertebrates, which occurs in several defined steps (Figure 1.22). Muscle forms as a result of the fusion of many individual muscle precursor cells called *myoblasts* which arise from mesodermal tissue. Myoblasts are already determined to become skeletal muscle; that is, they belong to a muscle **cell lineage**, even though they do not yet express the molecules that give skeletal muscle its characteristic contractile properties. Myoblasts are still able to divide, a property fully differentiated cells usually lose.

The differentiation of myoblasts into muscle depends on the expression of transcription factors belonging to the MyoD family (Figure 1.22b). These proteins are helix–turn–helix transcription factors (Book 1, Section 6.3.2) that

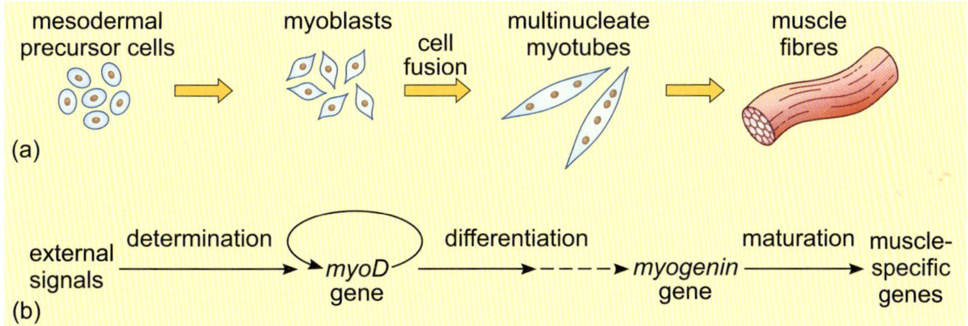

Figure 1.22   (a) Stages in the differentiation of skeletal muscle. Some mesodermal precursor cells develop into myoblasts, dividing cells that are determined to develop into muscle cells. Myoblasts fuse to form multinucleate myotubes. Groups of myotubes form a skeletal muscle fibre. (b) Some of the genes involved in differentiation of skeletal muscle. External signals (from other cells) stimulate mesodermal cells to develop into myoblasts. Myoblasts express genes such as *myoD* that encode transcription factors. The MyoD transcription factor stimulates further expression of its own gene, and promotes the further differentiation of the myoblasts so that they express other muscle-specific genes such as that encoding myogenin, another transcription factor. Ultimately, expression of muscle-specific genes, such as that encoding myosin, is stimulated.

bind to gene promoter and enhancer sequences, and act in a series to eventually activate transcription of the characteristic muscle-specific genes that encode contractile proteins, such as myosin II (Book 2, Section 5.3). These *myogenic* transcription factors also stimulate their own continued expression, making the switch to muscle protein expression permanent. The *myoD* gene product is one of the earliest-acting muscle cell fate determinants. In fact, when the *myoD* gene was inserted into a viral vector so that it was under the control of a constitutively active viral promoter (such that its expression was always 'on') and introduced into various other cell types (pigment cells, nerve cells, adipocytes, fibroblasts and liver cells), they too were converted into myoblast-like cells. Thus, *myoD* is required for *commitment* of cells to the muscle cell lineage. Other transcription factors of the MyoD family, such as myogenin, subsequently turn on other muscle-specific genes, resulting in further differentiation into skeletal muscle.

As well as switching on muscle-specific genes, *myoD* gene expression also inhibits the cell cycle by stimulating the transcription of p21 (Section 1.2.5).

■   How does p21 inhibit the cell cycle?

☐   It binds to and thereby inactivates cyclin–Cdk complexes.

At the same time, the myoblast withdraws from the cell cycle by down-regulating its growth factor receptors. The fibroblast growth factor FGF2 promotes myoblast cell division, and inhibits myoblast differentiation by suppressing the transcription of *myoD* genes, so its effect must be 'switched off' in order for the cell to differentiate. In fact, when myoblasts are grown in tissue culture, they can be induced to differentiate simply by removing all the growth factors from the growth medium. In this way, the fate of the myoblasts becomes *determined* (it can no longer be reversed or changed) because the expression of muscle-specific genes is permanently switched on and the cell stops dividing. Cell division and differentiation are thus mutually exclusive for this cell type.

This should give you some idea of the complex interaction between the cell cycle and cell differentiation, and their coordination by the expression of key transcriptional regulators. The key to understanding cell differentiation, therefore, is to understand what regulates the changing patterns of transcription factor activity.

1.4.4 Cell differentiation in mature organisms

The life histories of different types of cells in a multicellular organism are far from identical. Each kind of tissue has its own turnover time, often related to the function and workload of its cells. For example, the epithelial cells lining the gut form a barrier that is exposed to the gut contents, and is highly active in the absorption and secretion of molecules (Book 1, Section 2.4.5). These cells arise from a small population of constantly dividing cells at the base of the crypts (the valleys between the villi). As these cells divide, the daughter cells migrate up the villi, while differentiating further into the mature absorptive cells that express the appropriate transporters and enzymes. These cells live for only three to five days before they die and are shed into the gut lumen.

Epithelial cells in the protective skin epidermis are similarly recycled every two weeks or so, while the cells of an adult human liver have a relatively longer turnover time of 300 to 500 days. Some cells, including most neurons, never divide once they have differentiated into their specialised form, but survive, essentially in one position, sometimes for the entire life of the organism.

As development proceeds, the 'potency' of cell differentiation reduces and many cells become destined to produce only one or a few closely related types of cell. Most fully differentiated animal cells cannot normally dedifferentiate to become another type of cell; however, certain cells grown in cell culture will dedifferentiate when they are separated from their usual neighbours. For example, smooth muscle cells (such as those found in the walls of blood vessels) will dedifferentiate, ceasing to make smooth muscle myosin, if grown at low density in culture. This illustrates an important principle: the maintenance of a stable differentiated phenotype in many cases depends upon ongoing signals from neighbouring cells. Dedifferentiation can happen more readily in some organisms, including plants and amphibians, often as part of a regeneration process after damage. For example, during regeneration of an amputated newt limb, smooth muscle cells dedifferentiate and can give rise to cartilage cells as well as to new muscle cells.

So, if most differentiated cells in animals can no longer divide or change to give rise to new cells, how are the processes of renewal and repair maintained in mature organisms? Notable examples of cells that need to be replaced continually during the life of mammals are blood cells and epithelial cells of the skin and gut (see above). The continued production of large numbers of these cells is possible because of the existence, in mature animals, of populations of stem cells.

- What is the definition of a stem cell?

- A stem cell is an undifferentiated cell that divides asymmetrically to give rise to two different progeny: another stem cell, and a cell that will differentiate into a specialised cell type (Section 1.4.2).

There are several types of stem cell in animals. Embryonic stem cells derived from the inner blastula are totipotent, while cells from later stages of the embryo are pluripotent; that is, they can still develop into most cell types (Figure 1.23). Adult stem cells were, until recently, believed to be less flexible and only multipotent; that is, only able to give rise to a few different cell types restricted to certain cell lineages. Examples are haematopoietic stem cells that give rise to cells of the blood, epithelial stem cells of the intestine and of the skin, and neural stem cells that give rise to cells of the nervous system.

Recent evidence, however, indicates that we may have to revise our understanding of the nature of adult stem cells, and that some types may be much more flexible than previously thought. An example is that of neural stem cells, which have been reported to be able to differentiate into skeletal muscle cells in tissue culture when grown in certain conditions; and even *in vivo*, if first cultured in the appropriate conditions before transplantation (or grafting) into adult animals.

- Do cells of the nervous system and skeletal muscle share a common developmental origin?

- No, as shown in Figure 1.16, they are derived from different germ layers: skeletal muscle cells are derived from mesoderm, while nervous system cells are derived from ectoderm.

- Does the observation that skeletal muscle stem cells have the ability to develop into neurons in tissue culture suggest that this phenomenon normally occurs *in vivo*?

- No, the conditions in tissue culture may be very different from those *in vivo*, and neural stem cells may not normally be present in an area in which they would receive the signals needed to differentiate into skeletal muscle cells.

What determines when and how the progeny of stem cells in mature animals differentiate? As in earlier development, signals from other cells play an important role. The processes have been studied extensively in the haematopoietic system. In adult mammals, all blood cells are formed from stem cells that lie in the bone marrow. A range of growth factors, some of which are shown in Table 1.1, promote the differentiation of different types of blood cells.

Stem cells have the potential to produce unlimited numbers of cells in culture and consequently have become a major focus of regenerative medicine, which offers the prospect of growing cells and tissues to repair or replace

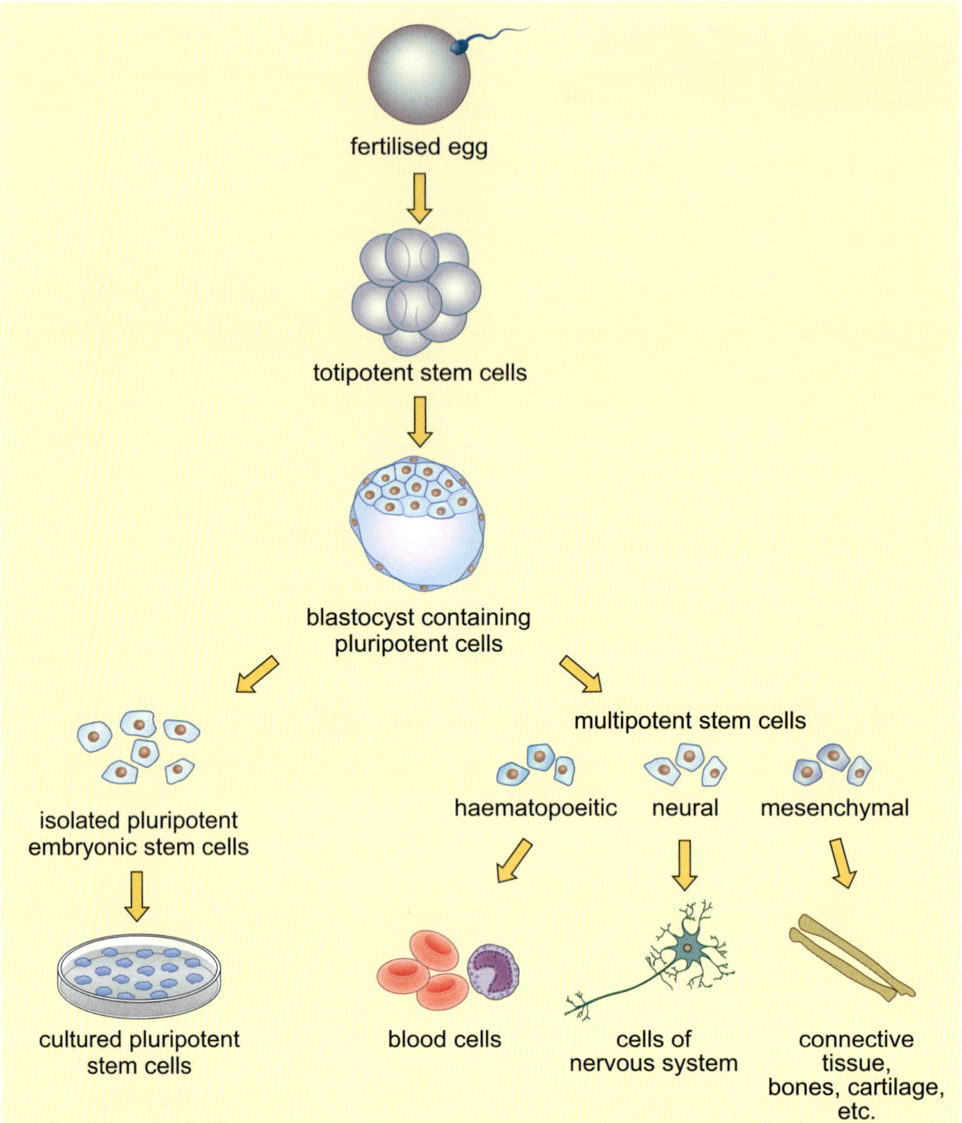

Figure 1.23 The hierarchy of mammalian stem cells and cell lineages. After fertilisation of the egg, the resulting zygote and the cells derived from the first few cleavages are totipotent; while those of the inner cell mass of the blastocyst are pluripotent (and are the source of so-called embryonic stem (ES) cells that can be cultured *in vitro*). Adult stem cells are generally more lineage-restricted (multipotent) and are generally referred to by their tissue of origin: for example, neural, haematopoietic (which give rise to blood cell types) and mesenchymal (which give rise to connective tissue cells such as osteoblasts and adipocytes).

non-functional or damaged tissues and organs. Adult stem cell treatments have, in fact, been successfully used for many decades to treat leukaemia and other related cancers by bone marrow transplants, and great deal of research is currently focused on working out how to reprogram stem cells by activating the signalling pathways that trigger differentiation into different cell types. You will learn more about some of these stem cell technologies in Chapter 4 of this book.

Summary of Section 1.4

- Cell differentiation occurs in multicellular eukaryotes as a result of differential gene expression. Gene expression is under combinatorial control, transcription of a gene depending upon the levels and types of transcription factors it is exposed to.

- The basic body plan of animals is predetermined by patterns of gene expression early on in development and cannot usually be changed by external influences, although some cell proliferation and differentiation carries on throughout life to maintain and repair tissues. In contrast, plants have much more developmental plasticity; they don't have a predetermined body plan, but continue to develop as they grow, in response to external factors such as light or nutrient availability.

- In very early development, unequal distribution of maternal molecules in the zygote can result in asymmetric cell division, in which different complements of maternal mRNAs or proteins are distributed to the two daughter cells. Many of these cell fate determinants are transcription factors, or mRNAs that encode transcription factors. This generates different cell lineages, which may have different developmental fates.

- Molecular gradients of extracellular signalling factors or other cell fate determinants can set up basic body patterns by acting directly on cells, which will respond differently depending on the level of the substance the cell is exposed to. Such molecules are known as morphogens, and overlapping morphogen gradients effectively provide cells with positional information.

- Cells also receive differential signals through interaction with their neighbours. These signals may be secreted extracellular factors such as growth factors, or cell–cell contacts. In both cases, intracellular signalling pathways are activated, which ultimately affect transcription. When such interaction results in a change in the developmental fate of a cell, it is known as induction.

- In some cases, cells become committed to follow a particular developmental fate some time before they express the genes that are characteristic of the fully differentiated cell. In a few cases, some cell types can be manipulated to dedifferentiate. The occurrence of such events demonstrates the need for ongoing signals to maintain differentiation.

- Differentiation of some types of cell, such as some epithelial cells and haematopoietic cells, continues throughout life. Such differentiated cells arise from populations of stem cells that are undifferentiated but may be partially committed cells. Stem cells divide by asymmetric division to form another stem cell and a cell that undergoes differentiation.

1.5 Final word

This chapter has illustrated how, in multicellular organisms, a large number of individual cell processes can be coordinated such that they occur in the correct sequence, at the correct time and location, to ensure that an organism develops its adult form. In animals, development is largely predetermined by a genetic programme, while plants have the ability to change their developmental pathway in response to external cues. You have learnt that cell proliferation, death and differentiation are coordinated by a relatively small number of highly conserved regulatory mechanisms, which include the cyclic activation of enzyme complexes (cyclin–Cdks) during cell division, the activation of proteolytic enzyme cascades in programmed cell death and the sequential activation of gene transcription during differentiation and development.

The next chapter explores the ability of populations of single-celled organisms to respond to their external environment and colonise a wide range of habitats.

1.6 Learning outcomes

1.1 Explain how progress through the cell cycle is regulated in prokaryotic and eukaryotic cells.

1.2 Describe the apoptotic pathway and distinguish between apoptosis and necrosis.

1.3 Outline the early stages of development in animal cells.

1.4 Describe and give examples of the mechanisms that promote different cell fates during differentiation and development, including asymmetric cell division, induction and morphogen gradients.

1.5 Outline the stages by which an undifferentiated cell becomes committed to a particular cell lineage.

Chapter 2 Microbial cells in the environment

2.1 Introduction

Microbes are present in large numbers in virtually all environments where water and nutrients are available. Soil, for example, is particularly rich in microbes, which are essential for the decomposition of organic matter and the recycling of nutrients. A gram of a soil might contain hundreds of millions of microbial cells belonging to hundreds of different species representing all of the major microbial groups, from prokaryotes (the Bacteria and Archaea) to eukaryotic fungi and protists. Artificial environments, including processed foods and synthetic substances, can also be colonised by microbes, with significant economic and health implications for human populations.

This chapter will look at the limits within which life is possible, and explore some of the adaptations that have allowed microbial cells to diversify and colonise almost all environments on Earth. To illustrate this, the chapter will consider the strategies and molecular specialisations that enable some microbes to inhabit extreme environments, ranging from arid deserts to the dark and cold conditions of the Antarctic seas. Finally, in Section 2.3, you will look at some examples of microbial interactions with other species that have also extended the range of environments microbes can inhabit.

Before reading this chapter, you should complete Activity 2.1; this is an internet-based activity in which you will identify the structural and biochemical features that characterise different microbial groups and which allow them to adapt to a range of environments.

Activity 2.1 Microbial diversity

(LO 2.1) Allow 1 hour

In this activity, you will use the internet to access websites containing information about the principal differences (structural, biochemical and ecological) between the main groups of microbes, notably bacteria, archaea, protists and fungi.

You will explore why certain types of cell, particularly archaea, are very good at colonising extreme environments.

You will participate in a group activity, directed by your tutor, in which you will record the results of your web search in a tutor group wiki, and discuss the activity within your tutor group.

In Book 2, Chapter 3 you learnt that organisms can use a range of different sources of carbon and energy to make the complex biomolecules (polysaccharides, lipids, proteins and nucleic acids) necessary for life. Autotrophs are able to synthesise complex organic molecules from simple inorganic compounds present in the environment using energy from light (photosynthesis) or inorganic chemical reactions (chemosynthesis). In contrast, heterotrophic animals and microbes cannot directly synthesise organic molecules from inorganic substances and require a supply of organic molecules synthesised by other organisms.

Autotrophs, mainly photosynthetic green plants and microbes but also chemotrophic microbes, are therefore the **primary producers** of organic material (or biomass) in an **ecosystem**; a community of living organisms interacting with their physical environment. Heterotrophs grow and produce biomass by metabolising carbon-based compounds derived from other organisms, and hence are known as the **secondary producers** in an ecosystem.

The availability of inorganic or organic nutrients is the main factor that determines the types of microbe present in an ecosystem. Microbes, particularly bacteria, respond rapidly to variations in nutrient levels, and during exponential growth, populations can expand enormously (Figure 1.2 in this book). As well as nutrient levels, a number of other key factors determine whether a microbe will flourish in a particular environment. These include temperature, pH (acidity or alkalinity), oxygen availability, salt concentration and atmospheric pressure. Most of the eukaryotic organisms with which you are familiar can only survive within a narrow range of conditions – most eukaryotic organisms for example can only tolerate temperatures up to about 50 °C, although some algae and fungi can tolerate up to 60 °C.

■ Why would very high or very low temperatures disrupt eukaryotic cell function?

☐ Their cellular components are only stable within a narrow range of temperatures. The structure of proteins, membranes and nucleic acids will be altered, or even denatured, above or below this range.

A number of eukaryotic organisms including fish, plants, invertebrates, and fungi have evolved strategies to cope with low temperatures and high pressures, but discoveries in recent decades have revealed that prokaryotic organisms can thrive in a much wider range of environments: for example, some aquatic bacteria can grow at temperatures in excess of 100 °C. Microbes living under extreme conditions are known as **extremophiles**, and some are **polyextremophiles** adapted for growth in at least two extreme conditions: for example, both high temperature and low pH. Extremophiles do not just tolerate the extreme conditions in which they live; they have an obligatory requirement for them. In fact, studying microbes collected from natural environments can be problematic because their growth requirements are often difficult or impossible to recreate in the laboratory (Box 2.1).

Box 2.1 Studying microbial activity in the environment

Relatively little is known about the number and distribution of microbial species and communities in natural habitats. About 30 000 named species of bacteria have been successfully cultured and studied in the laboratory, but this is a very small percentage (perhaps less than 0.1%)

of the many millions of bacterial species that probably exist on Earth. Many microbes have proved to be **non-culturable** when placed in standard laboratory culture media. This means they enter a dormant or 'resting' state in which they exhibit a low level of metabolic activity and often become very small. The cells remain viable (alive) but are unable to divide and grow. Indeed, the ability to enter this state is a strategy employed by many non-spore-forming bacteria in natural environments, and it enables survival under suboptimal growth conditions: for example, aquatic environments with low nutrient concentrations.

In addition, bacteria and archaea living in extreme environments usually have a highly specialised metabolism and it is difficult to replicate the conditions under which they can grow. It has therefore been necessary to develop other techniques for the study of the numbers, characteristics, and metabolic processes of non-culturable microbes. For example, to assess the presence of microbes in an environmental sample, a particular component of the microbial cells can be detected either directly or indirectly using labelling techniques.

- What specific component of the cell might you 'label' to identify and count methanogens in an anaerobic sediment sample containing a mixed population of microbes?

- Methanogenesis involves various specialised electron carriers not found in other kinds of bacteria (Book 2, Section 3.7.2). The cell membranes of methanogens, which are archaea, also contain unique lipid components (Book 2, Section 2.4.1). Either of these groups of molecules could be used as targets to attach 'labels' for the identification of methanogens.

One of the electron carriers unique to methanogens is coenzyme F_{420}, which naturally absorbs light at 420 nm and fluoresces blue–green, enabling the general identification of methanogens in mixed populations of bacteria and archaea from anaerobic samples. Identification of a particular species, however, requires a fluorescently labelled probe (Book 1, Figure 2.6) that targets a molecule specific to the organism. Very often identification of microbe relies upon detection of species-specific DNA sequences. These techniques take advantage of the properties of nucleic acids to form duplexes via the complementary base-pairing. You encountered this earlier in the module in the context of detecting mRNA transcripts (northern blotting, Book 1, Box 6.2), where a single stranded nucleic acid probe molecule can anneal to a target molecule immobilised on a membrane. The technique of *in situ* hybridisation uses sequence complementarity to visualise complementary sequences *in situ*, for example within a cell or on a chromosome. Nowadays, probe sequences are labelled with a fluorescent tag so that hybridised probe sequences bound to the cell or chromosome can be visualised easily by fluorescence microscopy (Book 1, Box 2.4). The diversity of microbes in a sample taken from an environment can be assessed visually by using a range of probes specific for different species

(Figure 2.1) and the intensity of fluorescence can be used to estimate the number of cells present.

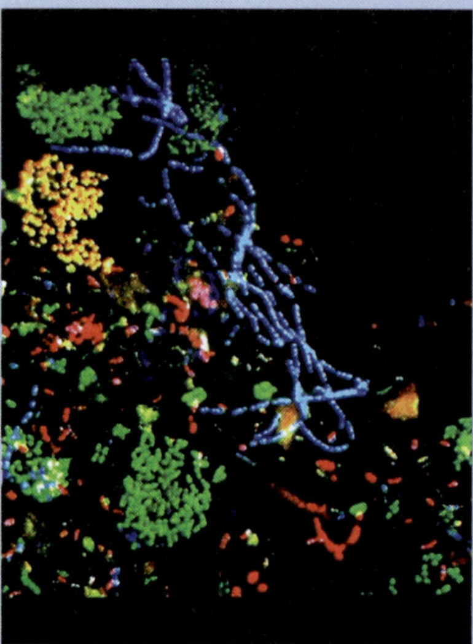

Figure 2.1 Use of multiple fluorescent-labelled DNA probes to assess microbial populations in a sewage sample. Different fluorescent probes, each targeting a specific group of bacteria, have been applied to the sample.

The next section explores the effects on cells of a range of environmental factors, and the cellular adaptations that allow microbes, particularly extremophiles, to inhabit certain environments. The study of extremophiles has attracted a great deal of interest in recent years. Phylogenetic relationships have suggested that the earliest cells on the Earth may have been extremophiles similar to present-day archaea, that may have had to withstand high temperatures, low oxygen availability, intense UV light exposure and extremes of pH and salt concentration (Book 1, Section 1.1). In the search for life on other planets, astrobiologists have turned to the study of extremophiles as models for organisms that may exist in liquid water elsewhere in the Solar System and beyond. Extremophiles are also a rich source of enzymes with unusual properties, such as acid or thermal stability, that are used in industrial and manufacturing processes.

Summary of Section 2.1

- Microbes are ubiquitous on Earth and grow in most natural and artificial environments.
- The study of microbes from natural environments requires specialised techniques. Specific 'labels' or probes can identify particular microbial species in samples from natural environments. Many microbes in natural environments appear to be non-culturable.

- Extremophiles are microbes that show specific molecular adaptations to extreme environmental conditions, such as high temperature and low pH. Polyextremophiles are adapted for life in multiple extremes.

2.2 The effect of environmental factors on microbial growth

Many factors, including nutrient availability, oxygen concentration, temperature, pressure, pH and salt concentration, can affect the growth of microbes, and most grow best within a narrow range of parameters. Environmental changes over time have imposed selection pressures that drive the evolution of new species that are able to colonise locations that may have been previously unavailable. Prokaryotes seem to be particularly adept at colonising a wide range of habitats; in fact bacteria and archaea comprise the great majority of known extremophiles.

While adaptations to extreme conditions often involve fixed inherited characteristics that have evolved as a result of natural selection (e.g. changes in the primary structure of essential proteins), other responses are transient and can change in response to fluctuating environmental conditions. Some of these variable responses involve the rapid modulation of gene expression. All cells respond to a sudden rise or fall in temperature with transient changes in gene expression, known respectively as the heat shock response and the cold shock response. A key part of these responses involves increases in the transcription of genes encoding stress proteins known as heat shock proteins (HSP) and cold shock proteins (CSP) respectively. Many HSPs are chaperones (Book 2, Section 1.5.3); these ensure that the cell's proteins remain correctly folded and active, or target terminally misfolded proteins to proteases for degradation. Some CSPs, on the other hand, seem to act as RNA chaperones, preventing the formation of secondary structures in mRNAs that might hinder gene expression. Another example of a variable response is the effect of temperature and pressure on cellular membranes, whose phospholipid composition can change quite dramatically in response to these factors.

There are, however, limits to this adaptability. Microbes that have, for example, evolved to live at very high temperatures are unable to grow at much lower temperatures. It should be emphasised that this is an expanding field of research; only a limited amount is known about the mechanisms underlying extremophile adaptation and why extremophiles can only survive within particular limits. This section will consider a range of environmental factors affecting microbial growth, beginning with oxygen, and explore some of the mechanism that are thought to contribute to the remarkable characteristics of some extremophiles.

2.2.1 Oxygen sensitivity

Most multicellular organisms on Earth live in an aerobic environment and require molecular oxygen (O_2) to carry out cellular respiration (Book 2, Section 3.6).

■ What is the role of O_2 in cellular respiration?

□ It acts as the final electron acceptor of the electron transport chain which captures energy in the form of ATP.

Many microbes are aerobes and either require oxygen, or grow better in its presence. In the laboratory, aerobic microbes can often be successfully cultured on the surface of agar plates, or in liquid cultures which are shaken rapidly to mix air with the liquid medium.

However, oxygen is also inherently dangerous to all cells. Cellular reactions involving oxygen generate toxic derivatives known as reactive oxygen species (ROS) (Book 1, Section 5.4.2), including the superoxide radical (O_2^-), hydrogen peroxide (H_2O_2) and the hydroxyl radical (•OH). These are generated as by-products of respiration and are highly reactive because they have unpaired electrons. All three are strong oxidising agents that can cause damage to cellular proteins, DNA and lipids. The hydroxyl radical is the most toxic of all, but it is transient and quickly removed by other reactions. Organisms that live in the presence of oxygen have, however, evolved specific mechanisms for removing superoxide and hydrogen peroxide. The enzyme superoxide dismutase (SOD) converts superoxide into hydrogen peroxide:

$$2O_2^- + 2H^+ \xrightarrow{\text{superoxide dismutase (SOD)}} H_2O_2 + O_2$$

The enzymes catalase and peroxidise convert hydrogen peroxide into water and oxygen:

$$2H_2O_2 \xrightarrow{\text{catalase}} 2H_2O + O_2$$

$$H_2O_2 + NADH + H^+ \xrightarrow{\text{peroxidase}} 2H_2O + NAD^+$$

The presence of these three enzymes is essential for aerobes to deal with toxic ROS. In contrast, many anaerobes express low levels of SOD. The relative SOD activity in selected organisms is shown in Table 2.1.

Table 2.1 Specific activities of superoxide dismutase (SOD) in selected Bacteria. Data from Hewitt and Morris (1975).

Species	Specific activity of SOD/μmol min^{-1} mg^{-1}
aerobe	
E. coli	44.0
anaerobes	
Chromatium species	0.6
Desulfovibrio desulfuricans	0.6
Clostridium pasteurianum	0.5

■ What is the significance of low levels of SOD in anaerobes?

□ When exposed to oxygen, their macromolecules may be subjected to damage by toxic ROS, notably superoxide.

The reasons why many anaerobes are very sensitive to oxygen is poorly understood, but vulnerability to toxic ROS may be a factor. A recent discovery, though, is that some anaerobic archaea, for example *Pyrococcus furiosus*, express another enzyme that removes superoxide, namely superoxide reductase (SOR), which requires the small iron–sulfur protein rubredoxin:

$$O_2^- + 2H^+ + \text{rubredoxin}_{\text{reduced}} \xrightarrow{\text{superoxide reductase (SOR)}} H_2O_2 + \text{rubredoxin}_{\text{oxidised}}$$

■ How does SOR differ from SOD?

□ SOR generates hydrogen peroxide but does not produce oxygen.

A peroxidase-like enzyme then breaks down hydrogen peroxide to yield water as the final product.

Anaerobes use alternatives to oxygen (e.g. sulfate, nitrate, or carbon dioxide) as the final electron acceptor in their electron transport chain (Book 2, Section 3.7.1). Microbial responses to oxygen are, however, much more diverse than is indicated by a simple split into aerobes and anaerobes. At one end of the spectrum are the **obligate aerobes**, which require oxygen for growth and cannot grow in its absence; and at the other are the **obligate anaerobes**, which require an environment that is completely free of oxygen. In between the two extremes there are many microbes that exhibit varying requirements for oxygen. Some are **facultative anaerobes**, which can grow either in the presence or absence of oxygen. An example here is the well-studied bacterium *E. coli*. This species can grow as a typical aerobic heterotroph using oxygen as the terminal electron acceptor, but under certain conditions, in the absence of oxygen, it can use nitrate instead. **Microaerophiles** are aerobic microbes that require low oxygen concentrations. A number of nitrogen-fixing bacteria (Section 2.3.3) belong to this category. They can only fix nitrogen under low oxygen concentrations; whereas, when using a source of fixed nitrogen such as nitrate, they behave as typical aerobes. Finally, some anaerobic microbes do not use oxygen, but are aerotolerant (i.e. they are capable of growing despite the presence of oxygen).

You have already met some examples of anaerobes in Book 2, Section 3.7, including the denitrifying bacteria and methanogens.

■ Where do methanogens obtain the carbon they use to synthesise organic molecules, and how do they generate ATP?

□ They use electrons from hydrogen to reduce inorganic carbon dioxide and produce methane (CH_4). The electrons are transferred through an electron transport chain to produce ATP.

All methanogens are archaea. They thrive in habitats where all electron acceptors other than carbon dioxide are absent. Methanogenesis is essential for the final stages in the decay of organic matter (Box 2.2) once other electron acceptors such as oxygen, sulfate, and nitrate have become depleted, and hydrogen and carbon dioxide have accumulated. Methanogens are important decomposers in sewage treatment. Methanogens are amongst the most fastidious obligate anaerobes known because their unique energy metabolism proteins, particularly the electron carrier coenzyme F_{420}, are inactivated in the presence of oxygen. Hence they are adapted to anaerobic condition and cannot grow in oxygen-rich environments. Special techniques are required in order to grow these archaea in the laboratory in the complete absence of oxygen.

Box 2.2 Microbes and decomposition

Ecosystems have many nutrient or biogeochemical cycles; complex pathways by which chemical elements or molecules move between the non-living (abiotic) and the living (biotic) components of the biosphere. Among the most important in terms of living organisms are the nitrogen cycle (Section 2.3.3) and the carbon cycle. Carbon is present in the atmosphere in the form of carbon dioxide (CO_2) which photosynthetic organisms, mainly plants, convert into organic molecules which are thereby made available to animals. Ultimately, carbon dioxide is recycled back into the environment by the processes of respiration and the decomposition of organic matter. Most decomposition occurs under aerobic conditions in soil, where dead organic material is broken down by the action of aerobic bacteria and fungi. These secrete enzymes that break down complex molecules such as plant cellulose to release carbon dioxide, ammonia and nutrients that are reused by living plants.

Anaerobic decomposition takes place in oxygen-free environments such as the animal gut (Section 2.3.2) or sediments at the bottom of stagnant lakes. It is the outcome of a complex sequence of microbial interactions, but can in general be simplified to a three-step process (Figure 2.2). Polymers such as plant cellulose and proteins are degraded to monomers by hydrolytic bacteria, and to a lesser extent by protists and anaerobic fungi (step 1). The monomers, notably sugars and amino acids, are used as substrates by a variety of fermentative bacteria (step 2). Major products here are acetate, H_2 and CO_2. In the final step 3, the specialised autotrophic metabolism of the anaerobic methanogens (Book 2, Section 3.7.2) converts acetate, H_2 and CO_2 into CH_4 and CO_2.

Anaerobic decomposition is usually the final stage in the treatment of sewage. Primary treatment to separate solids is aerobic and results in a clarified liquid effluent that can usually be discharged into rivers. Insoluble material from the primary stage (mainly cellulose and other fibrous polysaccharides) is collectively termed sludge. This is subjected to a secondary treatment stage under anaerobic conditions. Here, the microbial community breaks down the complex organic material to CH_4 and CO_2. The mixture of gases that emerges from the sludge digester is termed biogas, and can be used as a gaseous fuel.

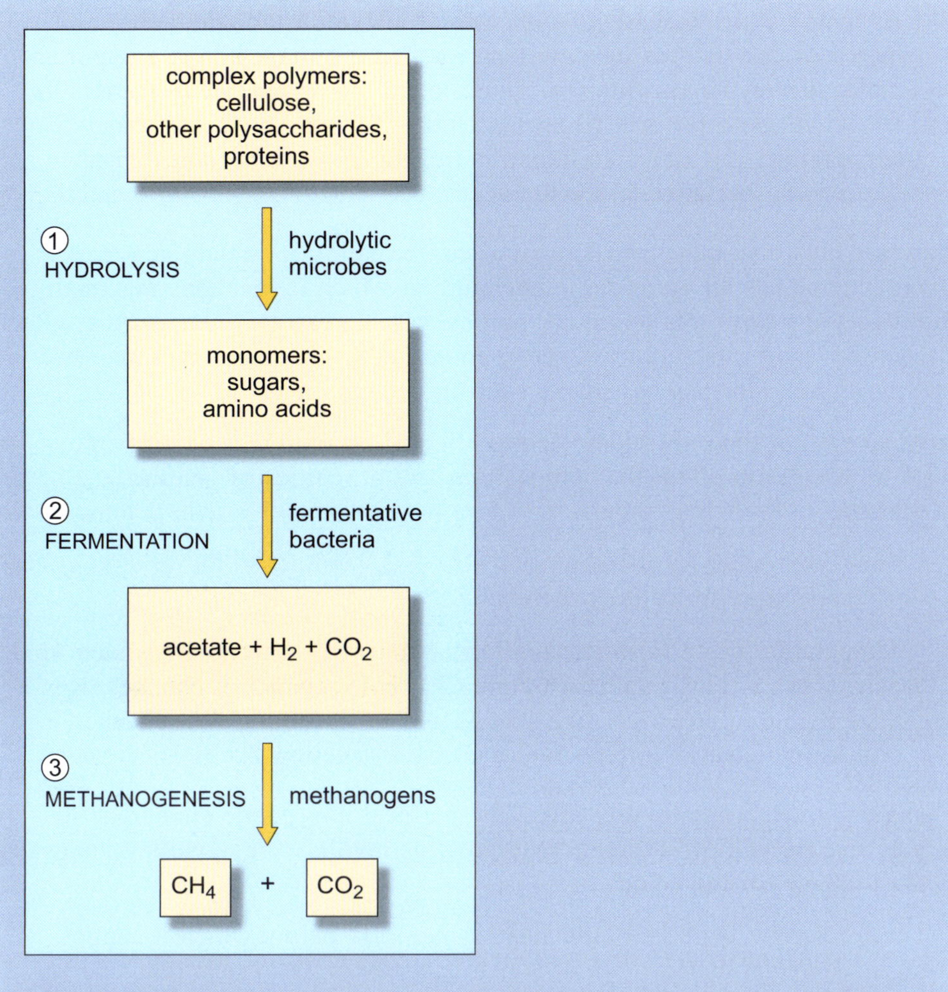

Figure 2.2 General scheme for decomposition under anaerobic conditions.

Most archaea are anaerobic, and while many live in familiar environments such as decaying organic matter, swampy land or the guts of animals, some live in extreme environments that cannot be tolerated by other organisms, including hot deep-sea vents and the cold, deep oceans. Temperature is perhaps the best-studied environmental factor in terms of its effects on cells and will be considered next.

2.2.2 Temperature

While microbes as a group show great diversity in their optimal growth temperature, no single species grows across a wide range of temperatures. Some have a very low optimum and may be active even below 0 °C, whereas others have a very high optimum and may grow at temperatures that exceed the boiling point of water at sea level.

■ From your study of enzymes in Book 2, summarise the general effects of increasing temperature on cellular enzyme activity.

☐ As temperature increases, so the rate of enzyme-catalysed reactions increases because the increased energy in the system allows more molecules of the enzyme's substrate to overcome the energy barrier for conversion into product (Book 2, Section 1.8.1). Above a certain temperature, the active conformation of the enzyme becomes unstable and the enzyme becomes less active.

Enzymes must maintain their structural conformation to remain active. A given enzyme has an optimum temperature at which it functions maximally, and deviations from this optimum, particularly increases in temperature, affect the non-covalent interactions between amino acid side chains that maintain the enzyme's three-dimensional shape, causing denaturation.

Proteins are not the only biomolecules affected by extremes of temperature. In Book 2, you learnt about the composition and properties of cellular membranes.

■ From your study of membranes in Book 2, summarise the general effects of temperature on cellular membranes.

☐ Temperature has a significant influence on membrane fluidity. Each lipid bilayer has a transition temperature (T_m) below which it changes from a fluid state to a solid, gel-like state. Membrane composition changes in response to changing temperature (Book 2, Section 2.5.3).

Changes in membrane fluidity affect the permeability of the membrane to solutes and the activity of membrane proteins involved in signalling pathways or the transport of molecules.

For each microbe there is a *minimum temperature* below which it cannot grow, an *optimum temperature* at which it grows at the fastest rate, and a *maximum temperature* above which it cannot grow (Figure 2.3). These three temperatures are called **cardinal temperatures** and are fixed, within very narrow limits, for each microbe. The minimum growth temperature may be determined by the temperature at which the cell membrane 'freezes' (i.e. becomes rigid), whereas the maximum growth temperature may represent the temperature at which key enzymes become thermally inactivated.

On the basis of their cardinal temperatures, microbes can be classified into four principal groups. Most microbes are **mesophiles**, which have mid-range temperature optima and occur in a wide range of terrestrial environments in temperate and tropical latitudes and also on, or inside, warm-blooded animals. *E. coli* is a typical mesophile, with an optimum growth temperature in rich culture medium of 39 °C, a minimum temperature of 8 °C and a maximum of 47 °C (Figure 2.4). **Psychrophiles** ('cold-loving' microbes) have very low temperature optima, **thermophiles** ('heat-loving' microbes) have high temperature optima and **hyperthermophiles** have very high temperature optima. The relationship between temperature and growth rates for selected examples in these categories is shown in Figure 2.4.

From Figure 2.4 you can see that *Polaromonas vacuolata*, a psychrophile, grows optimally at 4 °C; while in contrast, the hyperthermophile *Pyrolobus*

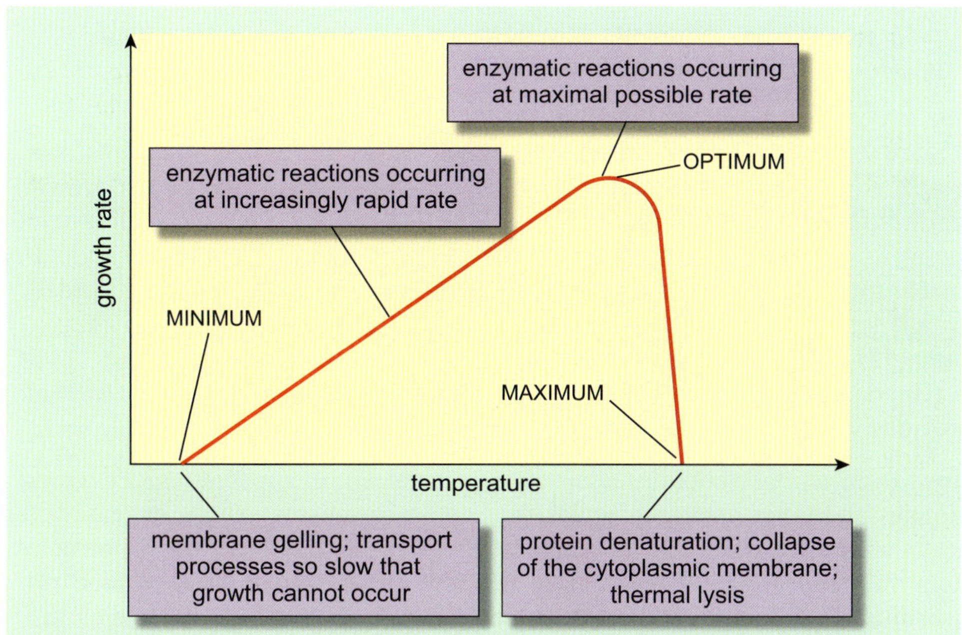

Figure 2.3 The effects of temperature on microbial growth rate; the three cardinal temperatures (minimum, optimum and maximum) are indicated.

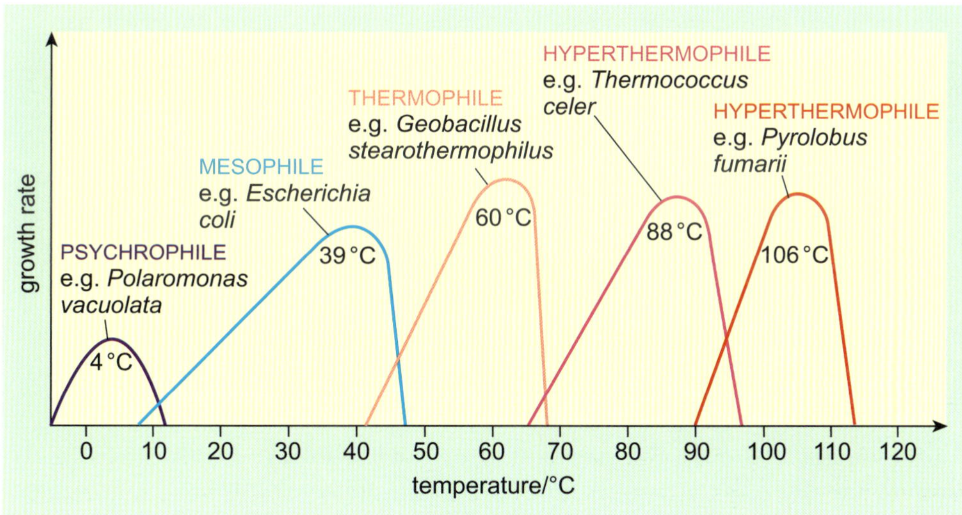

Figure 2.4 Relationship between temperature and growth rates for a typical psychrophile, mesophile, thermophile and two hyperthermophiles. The optimum temperatures for growth of each species are shown.

fumarii has a growth optimum of 106 °C. How are these microbes adapted for growth at the extremes of the temperature range?

Some like it hot – microbial life at high temperatures

Most of the very hot environments inhabited by microbes are associated with volcanic activity, notably the hot geothermal springs found in parts of Iceland, Japan, Central America and New Zealand, where the water is at or just below

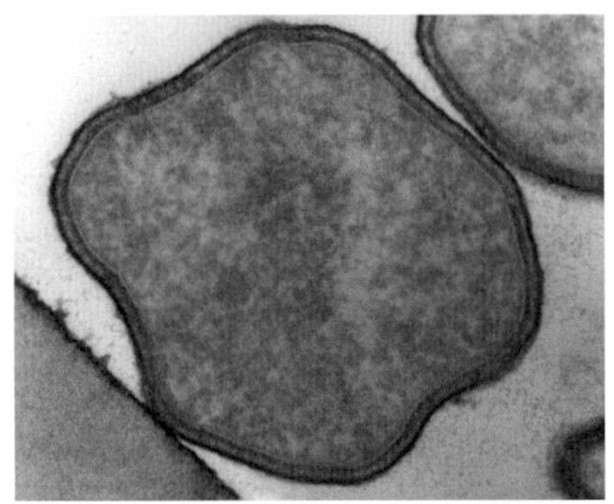

Figure 2.5 Electron micrograph of a thin section of *Pyrolobus fumarii*, one of the most hyperthermophilic of all known microbes (growth temperature optimum 106 °C).

boiling temperature. There are also deep-sea hydrothermal vents where temperatures can reach as high as 350 °C.

The majority of hyperthermophiles are archaea. *Pyrolobus fumarii* (Figure 2.5), one of the most hyperthermophilic of all known organisms, has a temperature maximum of 113 °C (Figure 2.4), well above the highest growth temperature known for a eukaryote (which is about 62 °C). *P. fumarii* lives in the walls of 'black smoker' hydrothermal vent chimneys (Section 2.3.4) and is unable to grow at temperatures below 90 °C. Currently, the highest recorded growth temperature for any organism is 121 °C for a strain (still known as strain 121) that was also isolated from a hydrothermal vent. Growth at 121 °C is remarkable because sterilisation in an autoclave at this temperature kills all previously described microbes.

How, then, are thermophiles adapted to grow at high temperatures? Enzymes, proteins and other macromolecules from these organisms exhibit increased heat stability (resistance to denaturation) when compared to similar molecules from mesophiles. Surprisingly, relatively few differences are seen in the structure of proteins in thermophiles compared to the equivalent proteins in mesophiles. Increased heat stability seems to result more from an increase in the number of non-covalent interactions between amino acid side chains.

The following case study considers malate dehydrogenase (MDH), one of the enzymes of the TCA cycle (Book 2, Figure 3.13). Three phototrophic bacteria were included in the study and their optimum growth temperatures are as follows: 32 °C for the mesophilic *Chlorobium vibrioforme*, 47 °C for the moderately thermophilic *Chlorobium tepidum* and 55 °C for the more thermophilic *Chloroflexus aurantiacus*.

MDH protein was isolated and purified from each of these bacteria and two thermal parameters were measured for each of the three purified proteins:

1 Half-life of thermal inactivation, which is the time taken in minutes for half of the protein to be thermally denatured at a specified temperature.

2 Transition temperature (or apparent melting temperature, T_m), which is a measure of the thermal stability of the protein and is the temperature at which the protein spontaneously denatures, i.e. unfolds.

This data, together with the optimum growth temperatures for the bacteria are shown in Table 2.2.

Table 2.2 Half-lives of thermal inactivation and transition (apparent melting) temperatures (T_m) for MDH from *C. vibrioforme* (*cv*-MDH), *C. tepidum* (*ct*-MDH) and *C. aurantiacus* (*ca*-MDH). Data from Dalhus et al. (2002).

Enzyme	Half-life (min) at		T_m (°C)	Optimum growth temperature of source microbe
	55 °C	65 °C		
cv-MDH	0.5	0	44.5	32
ct-MDH	>30	0	52.6	47
ca-MDH	>60	25	67.8	55

■   How does half-life of thermal inactivation compare for the three enzymes at the two temperatures studied (55 °C and 65 °C)?

□   For all three enzymes, the half-life of thermal inactivation is lower at the higher temperature, indicating an increased rate of thermal inactivation.

■   What is the relationship between the transition temperature (T_m) for MDH (i.e. thermal stability of this protein) and the optimal growth temperature for the microbe from which it was isolated?

□   The more thermophilic the bacterium, the higher the T_m of its MDH, so *ca*-MDH is the most thermally stable protein and *cv*-MDH the least thermally stable, while *ct*-MDH has intermediate thermal stability.

These data show that there is a correlation between the thermal stability of the MDH enzymes and the optimum growth temperature of the bacterium: the higher the optimum growth temperature, the more thermally stable the enzyme. What structural features of the MDH proteins account for this thermal stability? The study in which this data was reported (Dalhus et al., 2002) also investigated the primary structure of the three MDH enzymes and used X-ray diffraction (Book 2, Box 1.3) to investigate their three-dimensional structure (notably the tertiary and quaternary structure).

■   What feature of a protein is an essential requirement for studying it by X-ray crystallography?

□   The protein must be able to be crystallised.

As you can see from Table 2.2, the difference in optimum growth temperatures of the three bacteria is relatively small (32 to 55 °C); nonetheless, some structural features that contribute to thermal stability were identified.

In terms of quaternary structure, most MDH proteins previously studied were found to be dimers (i.e. consisting of two subunits). The three enzymes in this study were found to be tetramers, actually a dimer of dimers.

■   What types of interaction within the tertiary structure of each monomer could increase stability?

□   Hydrophobic interactions, ionic bonds, hydrogen bonds and disulfide bridges (Book 2, Figure 1.6) between different parts of the polypeptide chain help to stabilise the tertiary structure of the monomer.

Comparing *cv*-MDH from the mesophilic *C. vibrioforme* with *ct*-MDH from the moderately thermophilic *C. tepidum*, the greater thermal stability of the latter seems to be mainly accounted for by an increase in the number of polar amino acid side chains (Book 2, Table 1.1) forming additional hydrogen bonds within each monomer.

The *ca*-MDH from the most thermophilic bacterium in this study, *C. aurantiacus*, also has an increased number of hydrophobic interactions between aromatic amino acid side chains at the interface between the two subunits forming each dimer (Figure 2.6) and an increased number of hydrogen bonds and ionic bonds at the interface between the two dimers. Measurement of the volumes of the different MDH proteins suggests that the MDH protein with greatest thermostability has a more compact quaternary structure, reflecting the tighter interactions between subunits in the tetramer and tighter folding of the polypeptide in individual subunits.

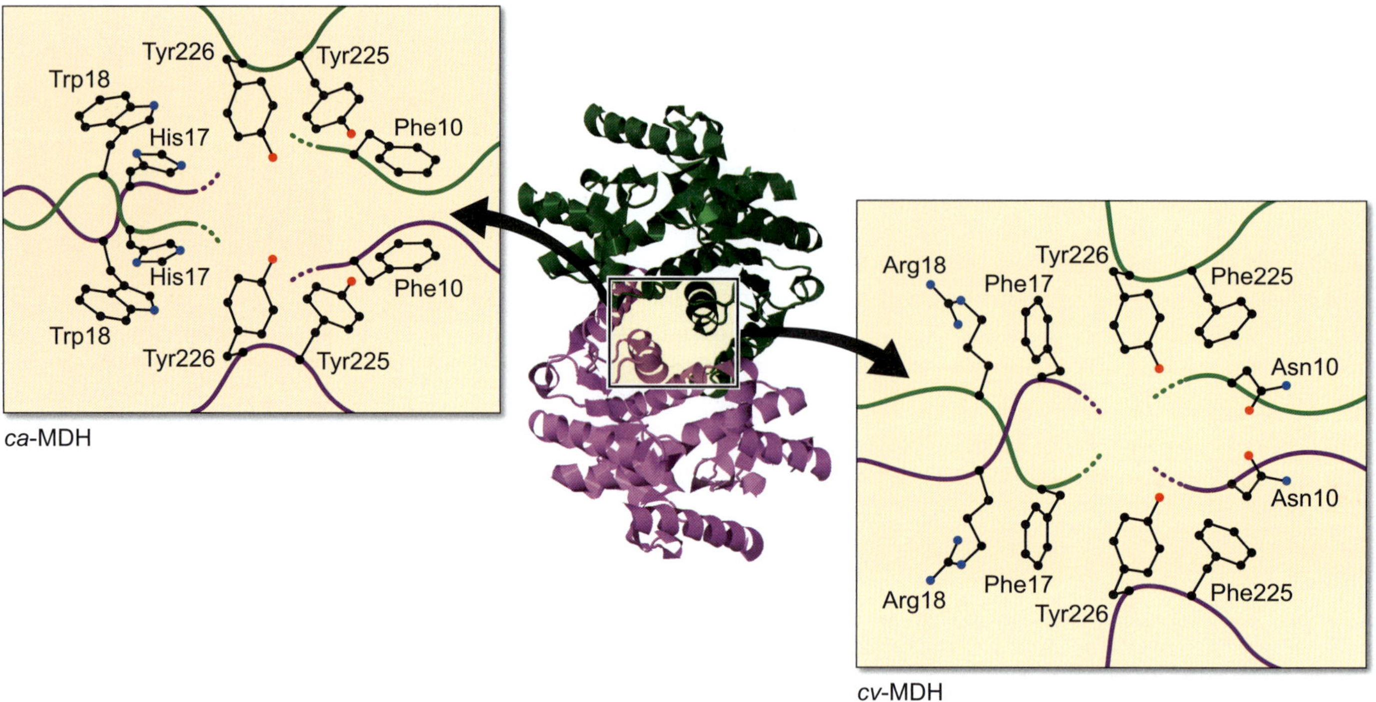

Figure 2.6 Central in the figure is a ribbon diagram of one of the dimers in *ca*-MDH with the two subunits coloured green and pink. The two detailed pictures show selected interactions at the interface between the two subunits forming the dimer in *ca*-MDH (left-hand panel) and *cv*-MDH (right-hand panel). The polypeptide backbones are colour-coded to match the ribbon diagram. Details of individual amino acid side chains are shown as 'ball and stick' representations. Notice especially the interactions between aromatic side chains His and Trp and Tyr and Phe. (Refer back to Book 2, Table 1.1 for further details of the amino acid side chains.)

■ What experimental technique could be employed to investigate the importance of individual amino acid residues for the thermal stability of *ca*-MDH?

□ Site-directed mutagenesis (SDM, Book 2, Section 1.9) could be performed to substitute an amino acid residue of interest with another residue with different chemical properties. The altered protein could then be expressed in bacteria, and its thermal stability investigated by determining half-life of thermal inactivation and T_m value in comparison to the wild type (non-mutated) protein.

Thermophile proteins in general tend to have a more tightly packed structure, with fewer loops extending from the surface, presumably helping to make them more resistant to denaturation at high temperature. The penalty for the rather rigid, thermostable structures formed by thermophile proteins is that they are much less active at low temperatures, probably because they are less able to make the small conformational changes required for catalytic reactions.

The term **extremozyme** has been applied to enzymes from thermophiles that function at high temperatures (although the term has been expanded to refer to enzymes from other extremophiles). The classic example is *Taq* DNA polymerase from *Thermus aquaticus*. This bacterium has an optimum temperature for growth of 70 °C, a maximum of 79 °C and a minimum of 40 °C. Originally isolated from thermal hot springs in Yellowstone National Park, it has also been isolated from artificial environments such as industrial and domestic hot water heaters, presumably contaminated from natural sources. *Taq* polymerase is stable at 95 °C and ideal for the PCR amplification procedure, which involves subjecting the enzyme to repeated cycles of low and high temperatures (Book 1, Box 5.1). A wide variety of extracellular enzymes such as amylases, proteases, and xylanases (enzymes that break down hemicellulose, one of the major components of plant cell walls) and intracellular enzymes such as dehydrogenases, oxidoreductases, and DNA polymerases have been isolated from hyperthermophiles. These enzymes, as well as being stable at high temperatures, are usually also highly stable against detergents, organic solvents and other chemical reagents.

Chaperone proteins are particularly important in thermophiles, where they function to maintain protein folding in an active conformation (Book 2, Section 1.5.3). In hyperthermophiles, the amount of chaperone protein per cell can increase significantly near the maximum growth temperature. For example, in *Pyrodictium* grown at 108 °C (2 °C below its upper limit), 80% of the soluble protein consists of the chaperone called *thermosome*, which maintains the other cellular proteins in a functional conformation.

Lipids and DNA from thermophiles also exhibit high thermal stability. The majority of hyperthermophiles are archaea, whose membranes are composed of phospholipids in which an ether linkage (rather than an ester linkage) joins the lipid tail to the glycerol component (Book 2, Section 2.4.1). Ether bonds are chemically more resistant, which contributes to the ability of archaea to survive extremes of temperature and pH. Archaeal membranes are also very rich in saturated fatty acids.

■ Why do saturated fatty acids protect membranes against high temperatures?

□ Membranes with a high proportion of saturated fatty acyl chains are less fluid than those with a high proportion of unsaturated fatty acyl chains. If the membrane was too fluid at high temperatures the cells may become fragile and vulnerable to physical damage (Book 2, Section 2.5.3).

A variety of DNA-binding proteins that may help to stabilise DNA at high temperatures have been isolated from thermophilic archaea. For example, a protein called Sac7d from *Sulfolobus acidocaldarius* binds non-specifically to the microbe's DNA and increases the temperature at which the two polynucleotide strands of the double helix denature (separate) by up to 40 °C. DNA-binding proteins like Sac7d may thus play key roles in stabilising DNA to heat denaturation.

Life in the freezer – microbial life in the cold

Most of the Earth's surface experiences temperatures below 15 °C, from the deep oceans to the soils, ice and snow of alpine and polar regions. Cold-adapted microbes are therefore more widespread than thermophiles. The coldest places on Earth are in the Antarctic, where large areas remain permanently frozen, or thaw for very short periods annually in the height of summer. Signy Island, in the South Orkneys, for example, has a mean air temperature range from 0.4 °C in January to −10.9 °C in July. Vast areas of ice form and melt on the surface of the Southern Ocean surrounding Antarctica, and here diverse communities of microalgae, mostly diatoms, have been identified (Figure 2.7).

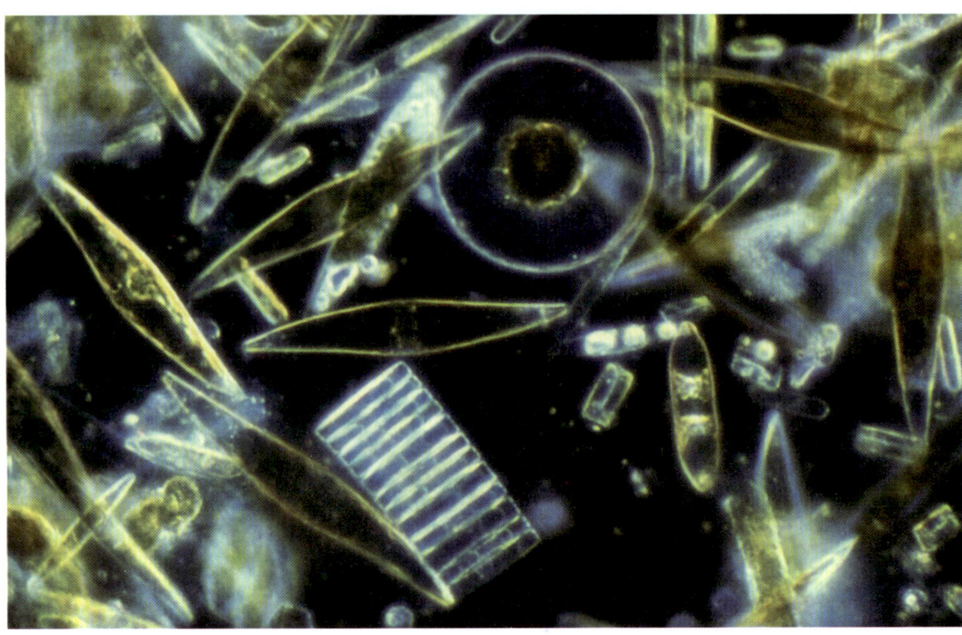

Figure 2.7 Light micrograph of phototrophic microbes taken from a core of frozen seawater from McMurdo Sound, Antarctica. Most of the organisms are either diatoms (as shown here) or green algae.

Figure 2.8 Endolithic community growing inside a sample of marble. The layers in the rock are as follows, from top down: a rust-stained surface layer receding into the background; a black lichen layer about 1 mm inside the rock; a light-coloured layer containing mainly fungi, but with some symbiotic algae; the innermost layer at a depth of about 8 mm, containing green algae and cyanobacteria.

Cold temperatures are often accompanied by other challenges such as high pressure in the deep ocean (Section 2.2.5), high salt concentrations in sea ice and intense UV irradiation and low availability of liquid water on the surface of ice or rocks. The green snow alga *Chlamydomonas nivalis* grows within lying snow as green-pigmented cells, which later produce spores containing bright red carotenoid pigment that create streaks of red or 'watermelon snow'. The carotenoid pigment protects the cells' chloroplasts from intense UV light, as well as absorbing heat from the Sun's rays, which melts the snow thereby providing the alga with liquid water.

In certain parts of Antarctica where there are exposed, translucent rocks such as marble and granite, communities of lithic (rock-dwelling) microbes are the dominant life forms. **Endolithic communities** (Figure 2.8) occupy the cracks inside the rocks, or the internal air spaces between the rock crystals (*endo-* means 'inside'). Life inside rocks has advantages in environments such as the Antarctic where temperatures and humidity are extremely low. The Sun heats the rocks and moisture from snow meltwater is absorbed. A wide diversity of microbes form these endolithic communities in which phototrophs are dominant.

How are microbes adapted to grow in cold environments? Enzymes and membranes tend to become more rigid when the temperature drops, reducing enzyme activity, membrane fluidity and permeability and the transport of nutrients and waste products. The primary response of microbes to low temperature is a slowing of all physiological processes. In addition, cold adaptation involves biochemical changes in cellular proteins and membranes, together with increases in the cellular concentrations of enzymes and other constituents to compensate for the depression of activity caused by the low temperatures.

Many cold-adapted microbes are able to maintain their membrane fluidity at low temperatures by changes in the fatty acid composition of their phospholipids (Book 2, Section 2.5.3).

- What types of membrane phospholipids are better suited to cold conditions?

- Unsaturated fatty acids with a short chain length help to maintain membrane fluidity at low temperatures.

An enzyme called desaturase catalyses desaturation of fatty acids, and decreases the ratio of saturated to unsaturated fatty acid chains in the bilayer. For example, the archaeon *Methanococcoides burtonii* is a cold-adapted methanogen from Ace Lake, Antarctica, where it lives in permanently cold (1–2 °C), methane-saturated water.

- How would you categorise this methanogen in terms of its temperature requirement?

- It is a psychrophile.

In an experiment to look at membrane composition, *M. burtonii* isolated from this environment was grown in culture at two different temperatures, 4 °C and 23 °C. The membrane phospholipids were extracted and analysed for the degree of unsaturation in two different classes of phospholipids. The greater the number of double bonds, the higher the degree of unsaturation. The results are shown in Table 2.3.

- Compare the degree of saturation of the two phospholipid classes at 23 °C and 4 °C in terms of the percentage with no double bonds.

- For both phospholipid classes, there is a higher percentage with no double bonds at 23 °C compared to 4 °C (84% versus 73% for phospholipid class 1; 74% versus 58% for phospholipid class 2).

Table 2.3 Estimated relative percentages of unsaturated archaeal phospholipids within two phospholipid classes from the methanogen *Methanococcoides burtonii* grown at 4 °C and 23 °C. Data from Nichols et al. (2004).

Number of double bonds	Phospholipid class 1 bond percentages		Phospholipid class 2 bond percentages	
	4 °C	**23 °C**	**4 °C**	**23 °C**
0	73.0	84.0	58.0	74.0
1	5.4	4.3	4.2	4.3
2	2.6	3.0	1.6	2.3
3	1.6	0.7	9.0	5.2
4	9.4	4.0	28.0	14.0
5	7.8	4.5	0	0
totals	99.8	100.5	100.8	99.8

■ For phospholipid class 2 only, compare the total percentage with one or more double bonds at both growth temperatures.

□ At 23 °C a total of c. 26% of the phospholipids have one or more double bonds, whereas this increases to c. 42% at 4 °C.

Overall, for both classes of phospholipids, the general pattern was the same: the lower the temperature, the higher the degree of unsaturation. In summary, cold adaptation in *M. burtonii* involves an increase in membrane lipid unsaturation which helps to maintain membrane fluidity.

Microbes inhabiting low temperature environments also have cold-adapted enzymes. An example is an extracellular amylase (which degrades starch) from a *Vibrio* species, isolated from sea sediment samples around Japan. The purified enzyme was shown to be inactivated at temperatures above 25 °C. This is probably a result of changes in protein structure that increase flexibility of the protein at low temperatures and so make it more easily denatured.

■ What types of changes in the protein might increase flexibility?

□ Reduction in the number of ionic bonds, hydrogen bonds or hydrophobic interactions formed between amino acid side chains.

Psychrophiles, then, are adapted to growth at low temperatures, but there is a lower limit: pure water freezes at 0 °C and seawater at −2.5 °C, and freezing stops microbial growth. Microbes show large differences in their ability to withstand freezing and there are several adaptive strategies to minimise freeze damage or even avoid it entirely. For example, Antarctic sea ice diatoms synthesise 'antifreeze' proteins that melt surrounding ice, making liquid water available, while cyanobacteria produce a gluey mass of extracellular glycoprotein (called mucilage) that prevents ice formation near cells. Increased solute concentration (e.g. sugars or glycerol) inside the cell can also reduce the freezing temperature of the intracellular fluid.

Cold-loving microbes are of interest to food processors who are interested in preventing food spoilage, and manufacturers who require cold-active enzymes, such as the proteases that are included in cold-wash laundry detergents. Lactose intolerance is widespread among human populations, and a psychrophilic β-galactosidase that can remove lactose during cold storage of milk was patented in 2012. Finally, psychrophiles also interest astrobiologists who have hypothesised that while most of the water on Mars is probably locked up in permafrost soils and the polar ice caps, some liquid water may be available where high salt concentrations depress the freezing point. If microbes have ever lived in this environment, they would probably have been polyextremophiles adapted to high salt concentrations and low temperatures.

2.2.3 Water availability and solute concentration

Liquid water is the solvent required for the function of all cellular biomolecules, so its availability is crucial to all living cells. The amount of 'free' water available for hydrating substances is known as the 'water activity' (a_w). When water interacts with solutes or surfaces, it becomes unavailable for other hydration interactions, so while pure water has a water activity of 1 (zero indicates an absence of any free water), the addition of solutes, such as salts or sugars, lowers the water activity. Microbes require a_w in the range 1.0 to 0.7. The a_w of seawater is 0.98; Great Salt Lake in Utah, USA has $a_w = 0.75$. Lowering the a_w in order to prevent bacterial growth is the basis for preservation of foods by drying (evaporation) or by addition of high concentrations of salt or sugar.

Common salt (NaCl) exits in a wide range of concentrations in natural environments. Organisms that grow best at low NaCl concentrations, but can grow in and tolerate NaCl are termed **halotolerant**. **Halophiles** (meaning 'salt-lovers') require sodium ions in order to grow. Most marine microbes are classed as mild halophiles (seawater contains about 3% w/v NaCl).

Extremely saline habitats are rather rare and most are in hot, dry areas of the world, such as Great Salt Lake, where the concentrations of the principal ions, Na^+ and Cl^-, are about 10 times those of seawater. The Dead Sea in Israel is rich in Mg^{2+}, in addition to Na^+ and Cl^-. These environments are home to extreme halophiles that require 15–30% w/v NaCl for optimum growth (Figure 2.9). Most microbes here are archaea (e.g. *Halobacterium*). Haloalkaliphiles such as *Natronobacterium* are extremely alkaliphilic as well as halophilic and grow in soda lakes (such as Mono Lake in California). Artificial saline environments include the surface of salted processed foods such as meat or fish.

Relatively few organisms have been able to adapt to highly saline environments. High solute concentrations normally dehydrate cells by drawing water out of their cytoplasm across the cell membrane by the process of osmosis (Book 2, Section 2.7.1). High salt concentrations inside the cell can also cause cellular proteins to aggregate together. Halophiles have two main strategies to prevent loss of water, both of which work by increasing the solute concentration inside the cell to achieve osmotic balance so that the cells do not dehydrate. The first strategy involves accumulating organic compounds such as sugars and amino acids which act as internal solutes. The second strategy, employed by extreme halophiles such as *Halobacterium*, is to

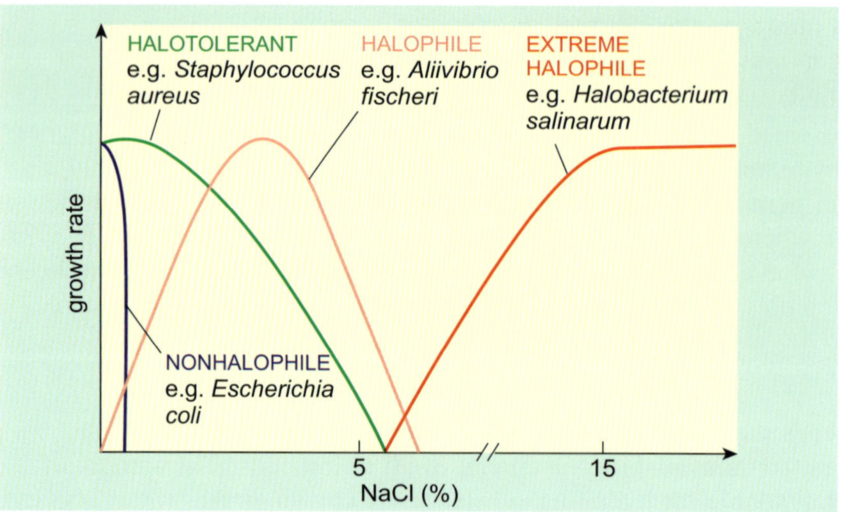

Figure 2.9 Effect of sodium chloride concentration on growth of microbes of different salt tolerances or requirements. The optimum NaCl concentration for marine microbes such as *Aliivibrio fischeri* is about 3%; for extreme halophiles, it is between 15 and 30%.

selectively pump Na^+ ions out of the cell at the same time as pumping in large amounts of K^+ ions from the environment. The proteins of these microbes are resistant to high ion concentrations. Many halophiles when placed in distilled water will immediately lyse (burst) due to the sudden change in osmotic conditions.

2.2.4 Microbes at extreme pH values

Cellular pH affects the state of ionisation of proteins and DNA. If the ionisation of amino acid side chains in proteins is altered (Book 2, Section 1.8.3), the ionic bonds that help to maintain the native conformation of the protein will be disrupted. Proteins may no longer be able to bind to ligands, enzymes may become inactive, and at extreme pH, proteins and DNA may be denatured.

Most natural environments on Earth have pH values between 6 and 8 and organisms growing under these conditions are termed **neutrophiles**. In contrast, **acidophiles** thrive in the rare habitats with pH values below 6 and cannot grow at neutral pH, whereas **alkaliphiles** require habitats with pH values above 8 (Figure 2.10). Some organisms, while not acidophiles, are acid-tolerant: for example, *Helicobacter pylori*, which can survive the very low pH of the human stomach where it is implicated in causing peptic ulcers. Highly acidic environments can result from geothermal activities, such as the production of sulfurous gases in hot springs or deep-sea hydrothermal vents and also in areas polluted by coal mining.

Acidophiles and acid-tolerant organisms maintain near-neutral intracellular pH values (between pH 6 and 8) while living in an acidic environments, by keeping the hydrogen ions (protons) out of the cell.

■   How might acidophiles remove hydrogen ions from inside the cell?

☐ Proton pumps in the cell membrane could remove hydrogen ions from the cytoplasm, thereby decreasing the intracellular concentration and raising the intracellular pH.

Most acidophilic bacteria, such as *Acidothiobacillus* species, and archaea, such as *Sulfolobus* species, have developed efficient proton pumps to maintain neutral intracellular pH. This is usually combined with membranes that are highly impermeable to protons, restricting their diffusion into the cell. The membranes of archaea such as *Picrophilus oshimae*, one of the most acidophilic microbes known, with a pH optimum of about 1, are particularly impermeable because of their ether-linked phospholipids. Some acidophiles, such as the bacterium *Acetobacter aceti*, maintain an acidic cytoplasm (with a pH as low as 4) and have evolved acid-stable proteins.

The most alkaliphilic microbes are cyanobacteria such as *Nostoc*, capable of growth at pH 12–13. There are also some alkaliphilic bacteria including some species of *Bacillus*. These organisms occur naturally in alkaline soda lakes. Alkaliphiles have the opposite problem to acidophiles. They need to pump hydrogen ions (H^+) across their cell membranes and into their cytoplasm to maintain a neutral internal pH. Some alkaliphilic bacteria have developed sodium transporters that drive the entry of sodium ions. A sodium–proton antiport (Book 2, Section 2.8.1) then exchanges the sodium ions that build up in the cytoplasm for protons from outside the cells, thus decreasing the overall pH of the cytoplasm (Figure 2.11).

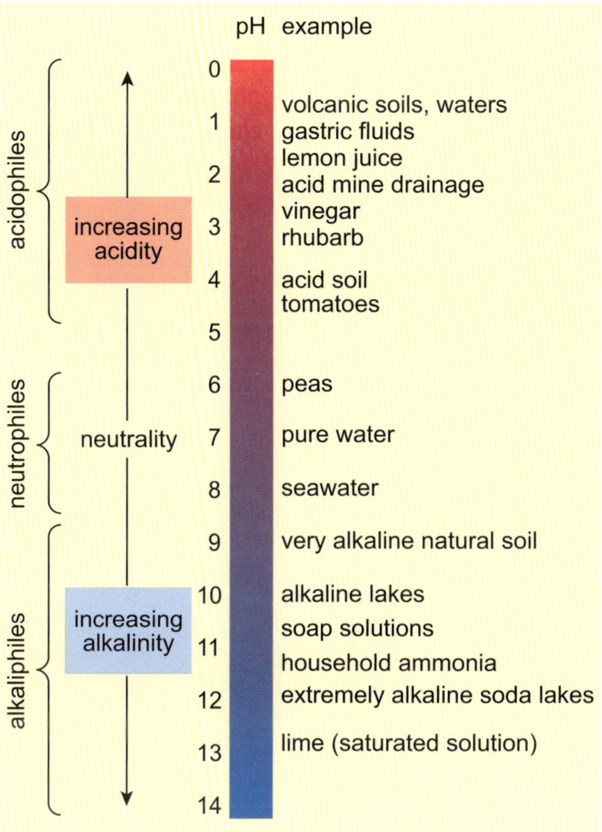

Figure 2.10 The pH scale. pH is a measure of the hydrogen ion (H^+) concentration. Solutions with a low pH value (a high H^+ concentration) are said to be acidic. Solutions with a high pH value (a low H^+ concentration) are said to be alkaline.

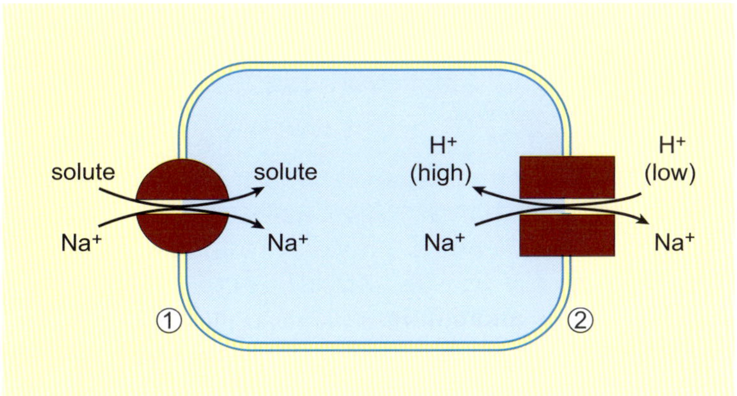

Figure 2.11 Mechanisms for pH homeostasis in alkaliphiles. (1) Na^+ is transported into the cell. (2) An antiport then removes Na^+ from the cell in exchange for H^+, thus lowering the cytoplasmic pH.

The chemiosmotic proton gradient is thus completely reversed in alkaliphiles (high inside the cell, low outside) compared to neutrophiles (low inside the cell, high outside). This presents a problem for the generation of ATP, which in bacteria is produced when protons cross the cell membrane by passing through the ATP synthase complexes to re-enter the bacterial cytoplasm (Book 2, Section 3.6.3). In alkaliphiles, the reversed proton gradient does not favour entry of protons through the ATPase. How alkaliphiles overcome this problem is unclear; nevertheless, they are perfectly able to generate ATP despite their alkaline environment.

Some organisms are polyextremophiles adapted for growth in at least two extreme conditions. Those requiring *both* high temperatures and low pH for growth are termed **thermoacidophiles**, while those requiring *both* high temperatures and high pH are termed **thermoalkaliphiles**. Figure 2.12 shows the distribution of a range of microbes as a matrix of temperature plotted against pH.

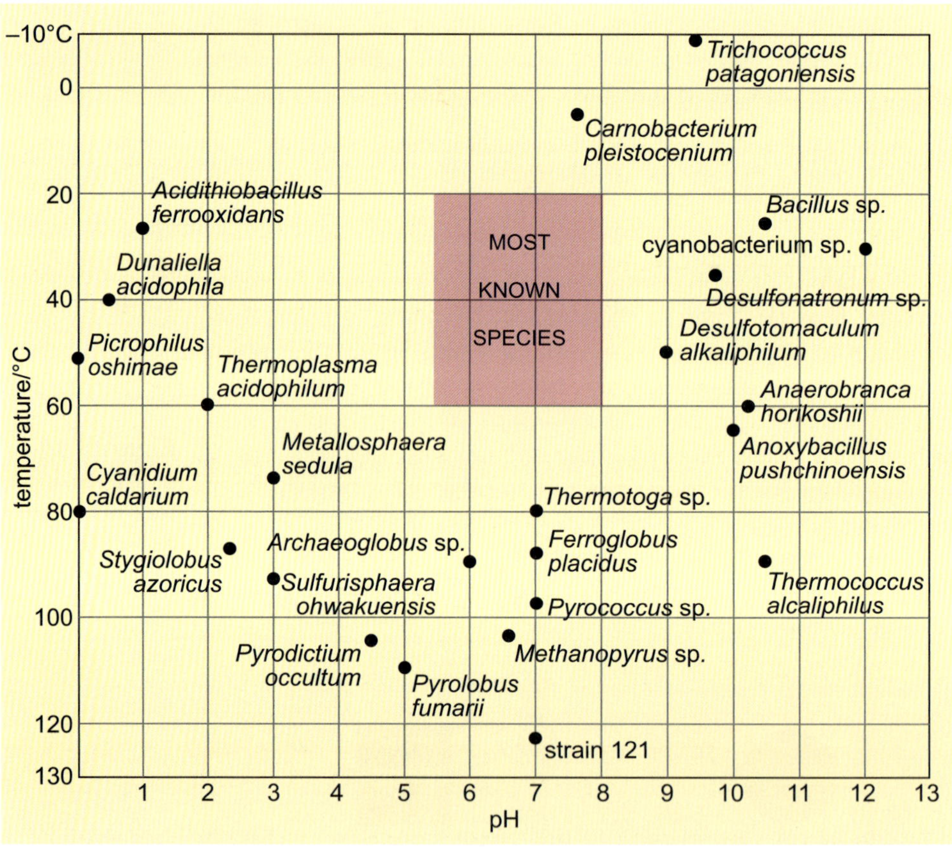

Figure 2.12 Distribution of a selection of known microbes in a matrix of temperature plotted against pH. The great majority of known microbial species are found in the pink boxed area.

- From Figure 2.12, identify the most extreme thermoacidophile and its growth conditions.

☐ *Cyanidium caldarium* (which is a red alga) is the most extreme thermoacidophile and grows at a pH of 0 and a temperature of 80 °C.

- From Figure 2.12, identify the most extreme thermoalkaliphile and its growth conditions.

☐ *Thermococcus alcaliphilus* is the most extreme thermoalkaliphile and grows at about pH 10.5 and 90 °C.

In fact, *Thermococcus alcaliphilus* is probably best classed as a hyperthermophile because it grows above 80 °C. It is also radioresistant (i.e. resistant to ionising radiation, Section 2.2.6), and hence has multiple extreme environmental requirements and is an exceptionally extreme polyextremophile!

■ From Figure 2.12, identify the pH requirement of strain 121.

☐ This strain grows at pH 7 and hence is a neutrophile; but recall that it is the most hyperthermophilic organism so far discovered.

2.2.5 Microbes at high pressure

Most of the volume of the Earth's oceans is at depths greater than 1000 m. Light only penetrates into the upper 100–200 m of the ocean, so no photosynthesis is possible in the deeper sea layers and the temperature is only 1–3 °C. Deep-sea microbes therefore tend to be chemoautotrophs and psychrophiles. The only other nutrients available tend to be falling organic matter produced in the higher photic zone. Organisms inhabiting the deep seas therefore encounter three environmental extremes: low nutrient concentrations, low temperatures and, as considered in this section, high pressure.

Pressure increases by one atmosphere (1 atm) for every 10 m depth of water, so a microbe growing at a depth of 5000 m must be able to withstand pressures of 500 atm. Some organisms simply tolerate high pressure but do not grow optimally under these conditions and are described as being **barotolerant**. They are usually unable to grow at pressure greater than 500 atm (Figure 2.13). Organisms that grow optimally under high pressure of 400–600 atm (although they can still grow very slowly at 1 atm, Figure 2.13) are termed **barophiles**.

The atmosphere (atm) is an international standard unit of pressure equal to the air pressure at sea level, and the pascal is the SI derived unit. It is equivalent to 101 325 pascals (Pa). One standard atm equals 101.325 kPa or approximately 0.1 MPa.

In deeper waters (below 10 000 m) extreme barophiles are present. For example, the bacterium *Moritella*, isolated from the Mariana Trench in the Pacific Ocean (>10 000 m depth), grows optimally at 700–800 atm. *Moritella* requires pressures greater than 400 atm in order to grow (Figure 2.13).

How are barophiles adapted to their extreme environments? High pressure tends to compress objects, and membrane lipids are particularly sensitive to the effects of pressure. Low temperature and high pressure both reduce the fluidity of cell membranes by increasing the tight packing of fatty acyl chains (Book 2, Section 2.5.3). This reduces the permeability of the membrane to water and other molecules.

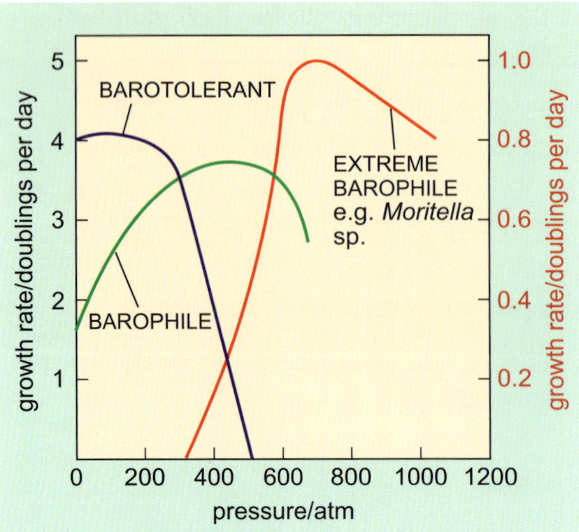

Figure 2.13 Growth of barotolerant, barophilic and extremely barophilic bacteria. Note that the growth rate of the extreme barophile *Moritella* (right-hand axis) is much slower than that of the barotolerant and barophilic bacteria (left-hand axis).

Barophilic bacteria modulate their membrane fluidity and composition in response to pressure. Figure 2.14 illustrates this with data from a study of the deep-sea bacterium *Photobacterium profundum*.

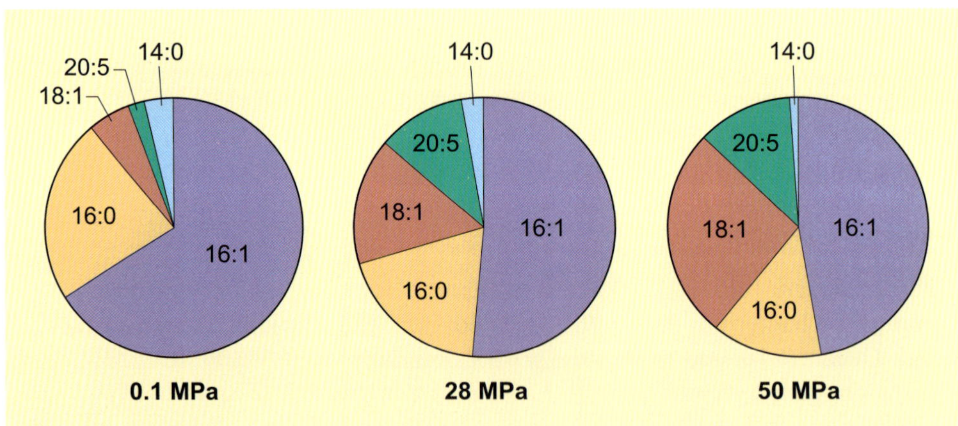

Figure 2.14 Effect of pressure on the proportion of the major fatty acyl species in the deep-sea bacterium *Photobacterium profundum*. The cells were grown under three different pressures and the relative proportions of different acyl chains were determined. The number before the colon is the carbon chain length; the number after the colon is the number of unsaturated C=C double bonds. For example, 14:0 is a saturated fatty acyl chain of 14 C atoms with no double bonds; 16:1 is an unsaturated 16-carbon fatty acyl chain with one double bond (i.e. mono-unsaturated); and 20:5 is a polyunsaturated 20-carbon chain with five double bonds. Data from Bartlett (2002).

- In Figure 2.14, how does the proportion of saturated fatty acyl chains change as a culture of the bacterium *Photobacterium profundum* is exposed to increasing pressure?

□   The proportion of saturated fatty acyl chains (both the 14:0 and the 16:0 chains) decreases with increasing pressure.

Organisms grown under high pressure have a higher proportion of unsaturated fatty acyl chains in their membranes which helps to increase membrane fluidity and permeability.

Pressure can also affect the level of expression of some genes. A specific outer membrane protein ompH (**o**uter **m**embrane **p**rotein **H**) is synthesised in cells grown under high pressure but not at 1 atm. OmpH is a porin (Book 2, Section 2.4.2), a channel protein that facilitates the diffusion of organic molecules through the outer membrane into the periplasmic space. Surprisingly few other proteins seem to be controlled by pressure, so adaptations to life in the deep sea may be relatively few.

2.2.6 Microbes and radiation

The final type of environmental extreme we will consider is exposure to ionising radiation. Gamma (γ) radiation is highly damaging to the biomolecules of all organisms and is widely used to sterilise medical equipment and in food preservation when the alternative method of heat treatment would be damaging to the material under treatment. However, some microbes, including the bacterium *Deinococcus radiodurans* and its relatives, exhibit the characteristic of **radioresistance**.

D. radiodurans was originally isolated from a tin of ground meat that had been exposed to 4000 Gy of gamma radiation, a dose that is roughly 250 times higher than is typically used to kill *E. coli*. A human would be killed by exposure to less than 5 Gy of ionising radiation. In fact, cells of *D. radiodurans* can survive exposure of up to 30 000 Gy of ionising radiation, a dose that would be expected to split its genome into small fragments. This bacterium has been isolated from near atomic reactors and other radiation sources that are lethal to most other organisms. Its name is derived from the Greek *deinos*, meaning 'strange' or 'unusual'.

The gray (abbreviated Gy) is a unit of radioactivity, used to denote the absorption of radioactive energy. One gray is the absorption of one joule of energy, in the form of ionising radiation, per kilogram of matter.

DNA has long been thought to be the principal target that determines loss of cell viability after exposure to gamma radiation.

■   Think back to what you have learnt about how cells cope with DNA damage (Book 1, Section 5.4). What might account for the radioresistance of *D. radiodurans*?

□   You might expect this organism to have an extremely efficient DNA repair mechanism that can repair damaged DNA even when the genome is in a fragmented state.

Several exceptionally efficient DNA repair pathways exist in these organisms for the rapid replacement of misincorporated bases and the repair of breaks in single- or double-stranded DNA (Book 1, Section 5.4.2). *D. radiodurans* also has multiple copies of its genome, which seems to help it to reconnect fragments of DNA and then repair them using homologous recombination. In addition, the nucleoids of stationary phase *D. radiodurans* cells have a unique arrangement of their DNA, which is condensed into a toroidal or ring-like structure, instead of being scattered within the cell (Figure 2.15). It is thought that this unusual DNA organisation also plays a role in radioresistance by assisting in DNA repair.

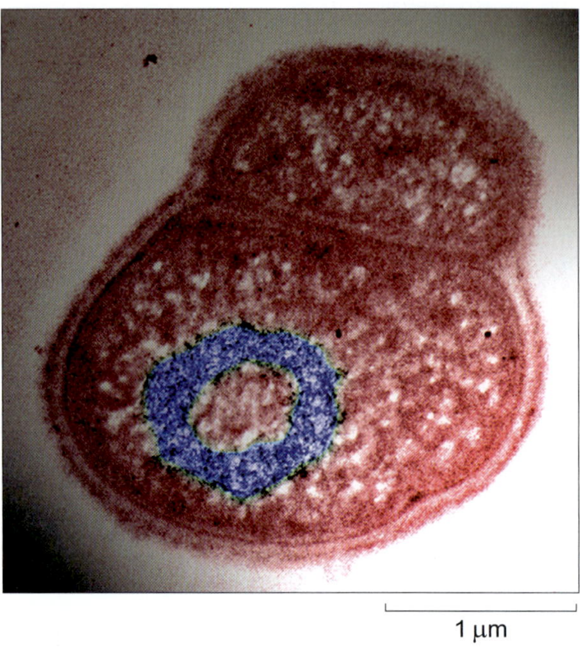

Figure 2.15 The radioresistant *Deinococcus radiodurans*. False colour transmission electron micrograph of a dividing cell coloured to show the toroidal morphology of the nucleoid (blue).

D. radiodurans is not unique in its radioresistance properties; some examples of other radioresistant species of bacteria are given in Table 2.4, with the relatively non-radioresistant *E. coli* included for comparison.

Table 2.4 Radioresistance in several examples of bacteria. D_{10} value defines the dose of gamma radiation needed to kill 90% of the irradiated population. Values for a given species vary considerably depending on growth conditions. Data from Cox and Battista (2005) and Trampuz et al. (2006).

Species	D_{10} value/Gy
E. coli	350
Acinetobacter radioresistens	2000
Chroococcidiopsis species (a cyanobacterium)	4000
Deinococcus radiodurans	10 000
Rubrobacter radiotolerans	11 000

■ Which species in Table 2.4 is the most radioresistant, and how does its radioresistance compare with that of *D. radiodurans*?

☐ *R. radiotolerans* has the highest D_{10} value, 10% higher than that of *D. radiodurans*.

Radioresistance is not restricted to the domain Bacteria: several thermophilic archaeal species also show extreme ionising-radiation resistance: for example, *Thermococcus gammatolerans*.

- What kind of extremophile is *T. gammatolerans*?

- This organism is a polyextremophile, tolerant of both high temperatures and high levels of gamma radiation.

D. radiodurans was originally isolated from an artificial, human-made environment; however, natural ionising radiation levels in most places on Earth are extremely low. This raises an interesting question: why should *D. radiodurans* have evolved such high levels of radiation resistance? In this context, it is interesting to note that *Deinococcus* species are polyextremophiles, resistant to a range of environmental challenges, including extreme temperatures, acidity and desiccation. The cellular mechanisms that enable *D. radiodurans* to cope with heavy damage to its genomic DNA may therefore be the consequence of environmental selection pressures other than radiation damage.

Alternative mechanisms for radioresistance have therefore been suggested: for example, it may be a side effect of a mechanism to cope with prolonged desiccation. Other researchers have suggested that proteins may actually be the most important target of ionising radiation, and that extreme radioresistance may have its explanation in the production of large amounts of antioxidant molecules that protect cellular proteins from damage by reactive oxygen species, such as superoxide (Section 2.2.1 above), which are generated by ionising radiation as well as many normal cellular processes. Protection of DNA repair proteins may in turn enable DNA repair pathways to work very efficiently. These alternative theories remain controversial, and in fact several mechanisms may have contributed to radiation resistance in these species. These interesting organisms are a good illustration of how studies on extremophiles continue to inform ideas and understanding in cell biology.

Summary of Section 2.2

- Microbes can be classified on the basis of their oxygen requirements as obligate aerobes, microaerophiles, facultative anaerobes or obligate anaerobes.

- Toxic reactive oxygen species (ROS) are challenging to all organisms. Aerobes have enzymes such as superoxide dismutase and catalase to remove ROS. Most anaerobes either lack these enzymes or exhibit low activity, and may be subject to damage in the presence of oxygen. Superoxide reductase is a novel enzyme found in anaerobic archaea that can remove superoxide.

- For each microbe, there is a minimum temperature below which it cannot grow, an optimum temperature at which it grows at the fastest rate, and a maximum temperature above which it does not grow.

- Psychrophiles have very low temperature optima, thermophiles and hyperthermophiles have high temperature optima, while mesophiles have mid-range optima.

- The upper temperature limit for eukaryotic growth is about 62 °C. Thermophiles and hyperthermophiles are therefore all bacteria or archaea, and are found for example in geothermal hot springs and deep-sea hydrothermal vents. The current upper limit for prokaryotic growth is 121 °C.

- Proteins, notably extremozymes, from thermophiles are heat-stable because of increased ionic bonds, hydrogen bonds and hydrophobic interactions in their tertiary and quaternary structures. Thermophile membrane phospholipids are rich in saturated fatty acids which prevents cell membranes becoming too fluid at high temperatures. Archaeal membranes contain ether-linked fatty acids, which are chemically more resistant than ester-linked fatty acids.

- In contrast, many microbes are able to maintain their membrane fluidity at low temperatures by increasing the proportion of phospholipids with unsaturated fatty acyl chains.

- Halophiles have a specific requirement for sodium ions. Extreme halophiles require 15–30% NaCl for optimal growth. Osmotic balance in extremely halophilic archaea is maintained by ion pumps.

- Acidophiles thrive in the habitats with a pH below 6, whereas alkaliphiles grow in habitats with a pH above 8. Proton transport in these microbes maintains intracellular pH values near neutrality, irrespective of the environmental pH. Neutrophiles grow between pH 6 and 8.

- Polyextremophiles grow under a range of multiple challenging environmental conditions. Thermoacidophiles grow at high temperature and low pH; thermoalkaliphiles grow at high temperature and high pH.

- Extreme barophiles require high pressures to grow optimally in the deep sea.

- Radioresistant *Deinococcus* is resistant to high levels of ionising radiation.

2.3 Microbial interactions

Many natural environments have low concentrations of the nutrients required for optimal microbial growth. The deep oceans, for example, are generally very **oligotrophic**, i.e. nutrient-poor. While the previous sections have concentrated on the cellular adaptations that allow individual microbes to occupy challenging habitats, some adaptations involve interactions with other organisms to provide an optimum environment. Many microbes grow either on the surfaces or within the bodies of other organisms, notably plants and animals (and occasionally other microbes), which are rich sources of both organic and inorganic nutrients. This final section of the chapter considers some examples of such interactions, especially from the perspective of nutrient supply.

Relationships with which you are probably familiar include:

- **Predation**: occurs when a heterotroph feeds on another organism. The feeding heterotroph benefits by gaining organic carbon.

- **Competition**: occurs when the populations of two species are both limited because of their joint dependence on a common nutrient.

Other relationships between species may be much closer and longer-term and are considered to be examples of symbiosis. Some interactions are mutually beneficial, and both partners gain something from the association. In some cases, the partners are so dependent on one another that they cannot exist in isolation. In other interactions, one partner benefits exclusively, sometimes to the detriment of the other. Symbiosis is a rather broad term, and many other often overlapping terms have been applied to classify different types of symbiotic relationship; here are listed some of those that are most relevant to microbial interactions, principally with animals, plants or other microbes:

- **Mutualism**: a symbiotic association between two organisms, from which both partners benefit. Mutualistic interactions range from those in which physical contact between partners is unnecessary, through to endosymbiosis (see below). The nature of the benefit to a particular partner has not always been clearly identified.

- **Endosymbiosis**: a very intimate association between two organisms, involving one organism living inside another. Structural and/or biochemical modifications of one or both partners that accommodate the other are a feature of these interactions.

- **Syntrophy**: also called *cross-feeding*; a specialised form of mutualism when the product of one species is used as the nutrient of another.

- **Commensalism**: a symbiotic association between two organisms, from which one partner benefits and there is a neutral effect on the other.

- **Parasitism**: a symbiotic association where one organism derives nutrients from the living cells or tissues of another.

- **Allelopathy**: occurs when one organism produces substances that influence the survival and growth of another type of organism. The effect may be beneficial or detrimental to the target organism.

There is not sufficient space here to cover the whole range of known microbial interactions; instead, this section considers selected case studies of microbe–microbe, microbe–animal and microbe–plant interactions to illustrate some forms of mutualism, competition, endosymbiosis and predation.

2.3.1 Predation: *Bdellovibrio*

There are numerous examples of prokaryotic and eukaryotic microbes, including many pathogens, that parasitise other eukaryotes. Microbes that parasitise or prey on other microbes are also common; many protists prey on bacteria, but prokaryote–prokaryote examples are rare. Here we consider an example of predation in which one bacterium preys on other living bacteria.

The curious bacteria of the genus *Bdellovibrio* uniquely invade other prokaryotic, usually Gram-negative, cells (*bdello-* means 'leech', from the mode of attachment to its host). The life cycle of *Bdellovibrio* is shown in Figure 2.16.

Each cell is small and comma-shaped, and is propelled at relatively high speeds by a single polar flagellum, making random violent attacks on a prey

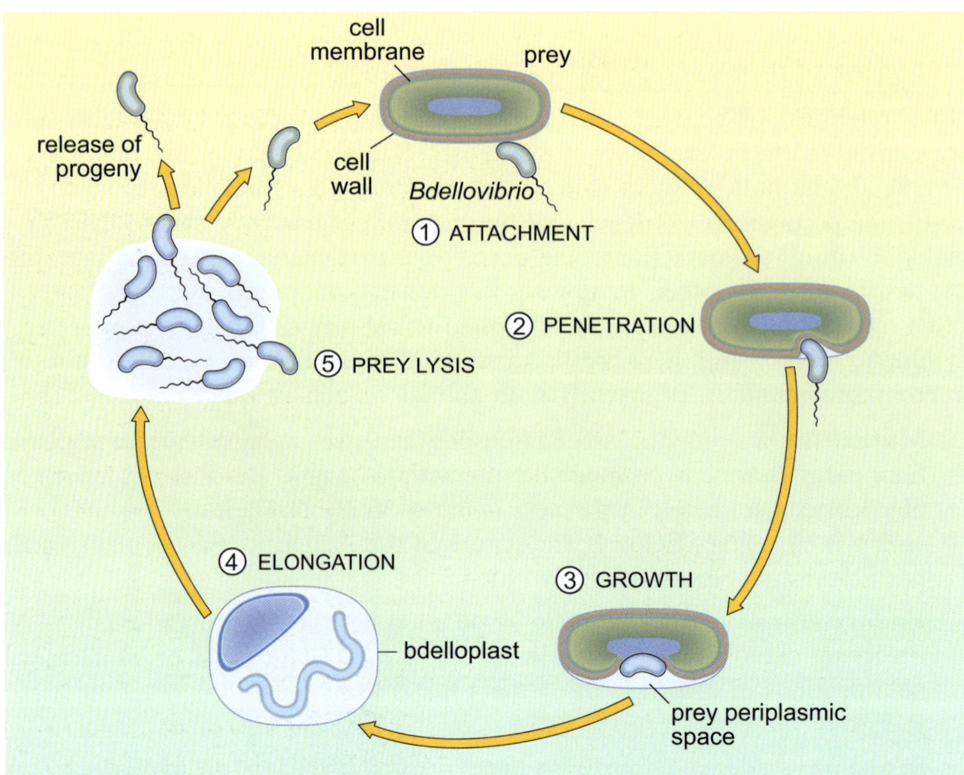

Figure 2.16 The life cycle of the bacterial predator *Bdellovibrio bacteriovorus*.

cell. The *Bdellovibrio* cell becomes attached end-on to the prey (Figure 2.16, stage 1). Entry is accompanied by rapid rotation to create a hole, and penetration is completed in 5–20 minutes (stage 2). Once inside, the *Bdellovibrio* grows in the periplasmic space between the prey cell membrane and the cell wall (stage 3). Here it elongates into a long spiral structure, inside what is termed the bdelloplast (stage 4). The elongated structure divides into motile daughter cells which are released, 3–4 hours after initial contact, by lysis of the residual cell membrane of the prey (stage 5). The number of progeny varies depending on the prey: *E. coli* usually produces about five, but up to 30 can be produced from other prey with larger cells, such as species of *Aquaspirillum*.

■ *Bdellovibrio* do not attack Gram-positive bacteria; suggest two reasons for this.

☐ (i) Perhaps *Bdellovibrio* are unable to penetrate the thick peptidoglycan layer that forms the cell wall of Gram-positive cells. (ii) Gram-positive cells lack the periplasmic space where *Bdellovibrio* grows in the prey bacteria (Book 1, Figure 3.6).

Bdellovibrio are obligate aerobes, widespread in soil, sewage and aquatic environments, including the sea. In addition to being predators, bdellovibrios are themselves attacked by bacteriophage. Other predatory bacteria have also been isolated and given names such as *Vampirococcus* and *Vampirovibrio*.

2.3.2 Mutualism and competition in the rumen

The rumen is the first section of the four-part stomach of certain mammalian herbivores known as ruminants, which include cows and sheep. Ruminants ingest tough plant material, which is softened in the rumen by the action of microbes, then regurgitated so that the semi-digested 'cud' can be re-chewed and broken down further. The relationship between the ruminant and the microbial populations of the rumen is an example of mutualism. The microbes in the rumen are provided with a constant supply of fermentable substrates. These microbes help to convert plant cellulose and other insoluble plant polysaccharides (Figure 2.2) mainly to short-chain fatty acids, which are thereby made available to the ruminant, along with proteins and sugars. Ruminants are thus able to survive on fibrous vegetation that most animals, including mammals, cannot digest. Both partners clearly derive benefits from the relationship.

In addition to their digestive function, rumen microbes also synthesise amino acids and vitamins for their own growth. Digestion of the microbes in the acid stomach releases these essential nutrients for use by the animal.

Environmental conditions within the rumen are anaerobic, with a relatively constant temperature (39 °C) and pH (6.5). The rumen contains a very wide range of microbes from the Bacteria, Archaea, protists and fungi. It contains particularly large numbers of prokaryotes (up to 10^{11} prokaryotic cells ml^{-1} of rumen fluid). The protist component of the ecosystem (up to 10^6 cells ml^{-1} of rumen fluid) is composed almost exclusively of ciliates, many of which are obligate anaerobes – a rare property amongst eukaryotes. The protists do not appear to be essential for rumen fermentation, but they do contribute to the overall process. Anaerobic fungi present in the rumen have also been shown to ferment cellulose.

Digestion in the rumen is a very complex process, but broadly follows the generalised scheme of anaerobic decomposition shown in Figure 2.2, but with some modifications (Figure 2.17). Food remains in the rumen for up to 12 hours. Anaerobic cellulose-degrading bacteria, ciliated protists and primitive fungi (called chytrids) hydrolyse cellulose into smaller units, mainly glucose, which then undergoes fermentation to produce short-chain fatty acids, primarily acetate, propionate and butyrate (2-, 3- and 4-carbon acids, respectively) and the gases carbon dioxide (CO_2) and hydrogen (H_2). The fatty acids are mostly taken up in the small intestine to become the main energy sources for the ruminant.

- What type of microbe is able to metabolise the gases (CO_2 and H_2) produced by fermentation in the rumen?

- Methanogens can convert them into methane (CH_4) (Figure 2.2).

The overall reaction of methanogenesis from carbon dioxide is:

$$4H_2 + CO_2 \rightarrow CH_4 + 2H_2O$$

Methane is released into the atmosphere and makes a significant input to so-called greenhouse gas emissions.

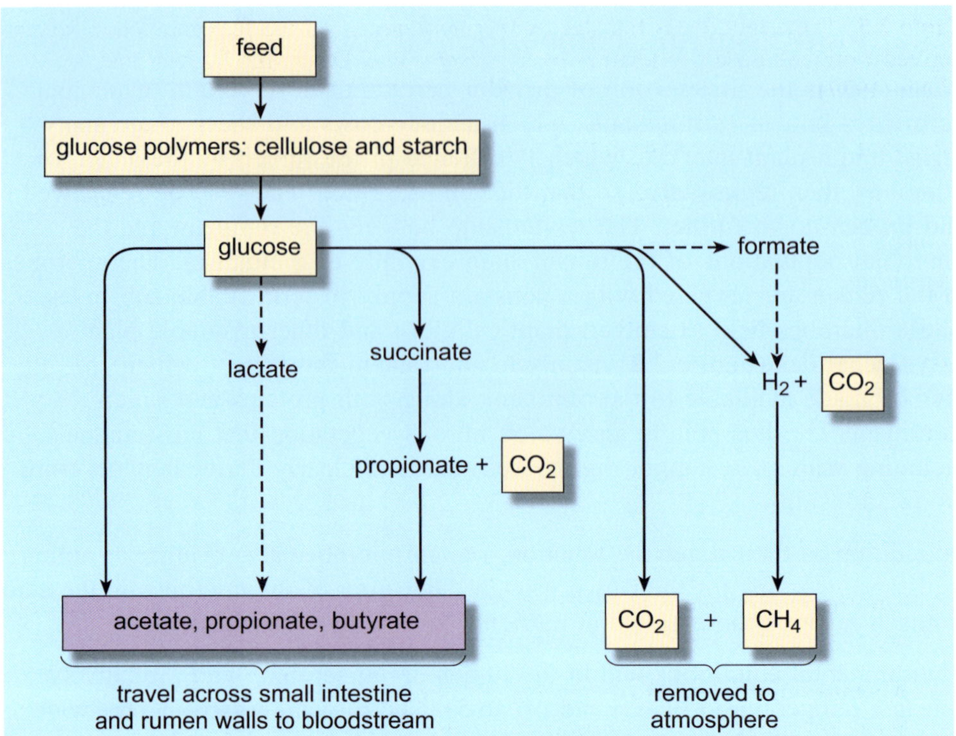

Figure 2.17 Biochemical reactions in the rumen.

The overall fermentation that occurs in the rumen (Figure 2.17) can be described by the equation:

$$57.5 \text{ glucose} \rightarrow 65 \text{ acetate} + 20 \text{ propionate} + 15 \text{ butyrate}$$
$$+ 60CO_2 + 35CH_4 + 25H_2O$$

In addition to mutualism, the rumen illustrates another important type of interaction: syntrophy, or cross-feeding (Table 2.5). Despite a potential daily production of 800 litres of H_2 gas by a 500 kg cow, the partial pressure (concentration) of H_2 gas in the rumen is actually very low at about 30 Pa. This is because the H_2 produced by fermentative organisms such as *Ruminococcus flavefaciens* is consumed during methane production by methanogens such as *Methanobrevibacter ruminantium*. H_2-using species need not necessarily be methanogens – other rumen bacteria are also H_2 consumers. This 'H_2 transfer' between two microbial species is an example of syntrophy, as illustrated by the data in Table 2.5, which shows that the products of fermentation by *R. flavefaciens* change in the presence of a methanogen.

Table 2.5 The effect of H_2 transfer to *M. ruminantium* on the fermentation products derived from cellulose degradation by *R. flavefaciens*. Data from Latham and Wolin (1977).

Product	mol l^{-1}/100 mol l^{-1} glucose units fermented to products	
	R. flavefaciens **alone**	*R. flavefaciens* + *M. ruminantium*
acetate	107	189
succinate	93	11
formate*	62	1
H_2	37	0
CH_4	0	83
CO_2	0	94

* A 1-carbon acid.

■ Summarise the main changes in the fermentation products that result from the syntrophic association between the H_2 producer and H_2 consumer.

☐ Acetate, methane and carbon dioxide become the principal products. H_2 is no longer an end product as it has been consumed during methanogenesis.

Degradation of cellulose in the rumen is not the sole province of bacteria: anaerobic fungi and ciliated protists are also involved. Like the bacteria, the fungi secrete cellulases and then take up soluble substrates. Here there is competition for available sources. The rumen fungus *Neocallimastix frontalis* produces substances that inhibit the cellulose-degrading activity of the bacterium *R. flavefaciens*, so this interaction is an example of an allelopathy, one that is detrimental to the host organism. The anaerobic bacteria and fungi are also in competition with cellulose-degrading ciliated protists, which are all phagocytic, ingesting small particles of the plant material that enters the rumen.

The rumen therefore illustrates a very complex range of different interactions:

* *mutualism* between ruminant and microbes
* *competition* (including *allelopathy*) and *syntrophy* are the two principal determinants of microbial community structure in the anaerobic rumen
* *predation* is also involved, because many of the rumen protists are active predators of the rumen bacteria.

The end products of rumen digestion that are not used by the ruminant are the gases CH_4 and CO_2. These gases accumulate in large quantities and are released to the atmosphere. Approximately 10^{14} g of CH_4 per year are produced by wild and domesticated ruminants, which represents about a third of all biologically generated methane; methanogenic activity in natural wetlands, paddy fields and by termites is the other major source.

2.3.3 Endosymbiosis and cooperation between specialised cells: nitrogen fixation

The element nitrogen undergoes a series of transformations on Earth through the activities of living organisms which collectively form the nitrogen cycle. **Nitrogen fixation** is a major component of this cycle, and is the process by which nitrogen gas (N_2) is reduced to ammonia (NH_3). It is of huge global significance because nitrogen is essential for the biosynthesis of biomolecules, including nucleotides and amino acids, in all living organisms. Nitrogen gas (N_2) is the most abundant element in Earth's atmosphere (comprising 80% of the atmosphere), but the strong triple bond between the nitrogen atoms makes it relatively inert; it doesn't readily undergo reactions with other molecules and is thus unavailable to most organisms. Only nitrogen-fixing prokaryotes are able to 'fix' N_2 gas by converting it into ammonia. Plants can then take up this nitrogen from the soil as ammonium (NH_4^+) or nitrate (NO_3^-) ions, and the nitrogen-containing biomolecules they synthesise become available to heterotrophic animals and microbes.

Only about 100 species of microbes can fix nitrogen (mostly bacteria but also a few archaea) and yet nitrogen fixation ranks with photosynthesis as one of the key processes that sustains ecosystems on Earth. Nitrogen fixation is carried out by a large microbial enzyme complex called nitrogenase composed of two metalloenzymes: the iron (Fe) protein and the molybdenum–iron (Mo–Fe) protein, which both contain iron–sulfur (Fe–S) groups. The overall reaction for nitrogen fixation can be summarised as:

$$N_2 + 8\,e^- + 8H^+ + 16ATP \rightarrow 2NH_3 + H_2 + 16ADP + 16P_i$$

The enzymatic reduction of N_2 to ammonia requires a very large input of energy in the form of ATP to drive the transfer of electrons from the Fe protein to the Mo–Fe protein, which ultimately breaks the covalent bonds between the nitrogen atoms. The electrons derive from a strong reducing agent such as reduced ferredoxin or NADPH, which also requires a great deal of energy to synthesise. Nitrogenase also works very slowly, so that nitrogen-fixing bacteria need to synthesise very large amounts of the enzyme (up to 10% of total cell protein) in order to achieve reasonable rates of N_2 fixation. (Look back at Book 2, Figure 1.14: the three bands with M_r identified are the subunits of the nitrogenase complex isolated from the free-living nitrogen fixer *Azotobacter chroococcum*. You can see that nitrogen fixation is energetically a very expensive process.)

Perhaps the best-known nitrogen fixers are rhizobia (e.g. *Rhizobium* and *Bradyrhizobium* species), Gram-negative bacteria that live symbiotically with plants in root nodules. It is easy to see the benefit that plants derive from this symbiotic relationship: a ready-made supply of usable nitrogen, but in what way do the nitrogen-fixing bacteria gain? Nitrogenase is strongly inhibited by oxygen (which reacts with the iron component of the nitrogenase proteins) and while this is not a problem for anaerobic nitrogen fixers, aerobic nitrogen fixers have had to develop strategies for protecting the process of nitrogen fixation from oxygen. Different microbes have solved the problem in various ways.

The association between rhizobia bacteria such as *Rhizobium* and *Bradyrhizobium* species and their host plants provides a special protective environment in the form of **root nodules**. In these endosymbiotic associations, the nitrogen-fixing bacteria provide usable nitrogen to their host and receive carbon substrates from the host, but of interest here are the mechanisms that maintain low O_2 levels within nodules. Rhizobia living in the soil infect the roots of plants of the legume family (family Fabaceae, which includes peas and beans), inducing the formation of root nodules (Figure 2.18).

Inside the infected root nodule cells, the bacteria are enveloped in a membrane derived from the plant cell membrane called the peribacteroid membrane (PBM). Here they divide and differentiate to form nitrogen-fixing bacteroids. The organelle-like structure composed of PBM and bacteroids is called the symbiosome and is the basic nitrogen-fixing unit of the nodule (Figure 2.19).

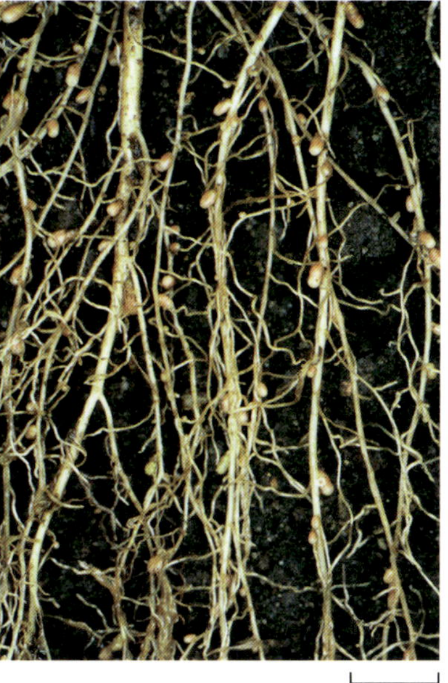

Figure 2.18 Nodules containing nitrogen-fixing *Rhizobium* on the roots of clover (*Trifolium* species).

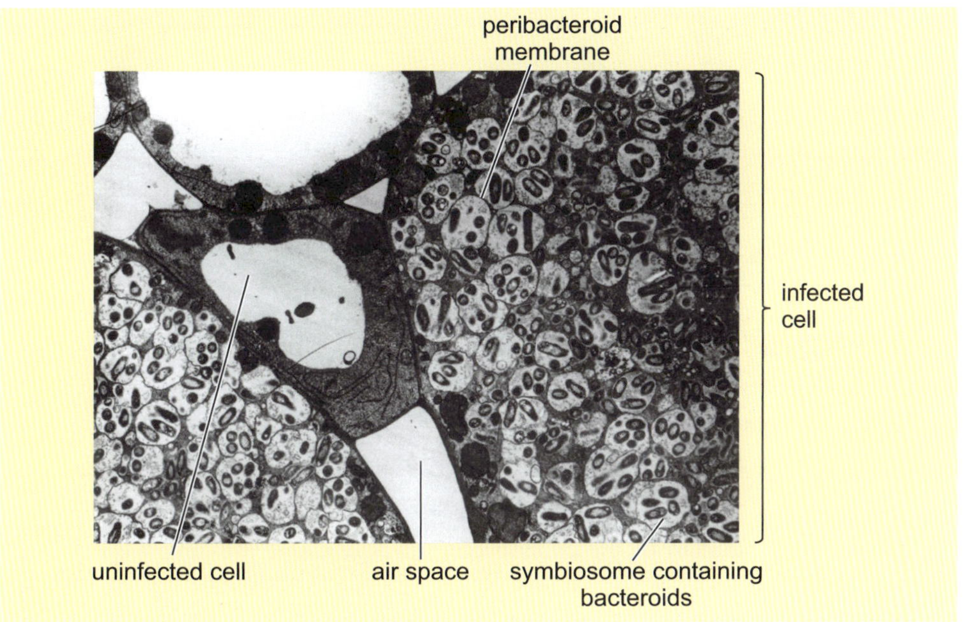

Figure 2.19 Electron micrograph showing cells in the infected zone of a soya bean root nodule. The infected cells containing large numbers of symbiosomes are clearly seen on either side of an uninfected cell and an air space. The peribacteroid membrane of a symbiosome is indicated in the right-hand infected cell.

Legume root nodules are pink in colour because they produce a haem protein called **leghaemoglobin** which binds oxygen, much as does the haemoglobin in vertebrate blood, facilitating diffusion of O_2 through root nodule cells and supplying O_2 to bacteria. The flux of O_2 released from leghaemoglobin exactly matches respiratory demand by the bacteria, so that plenty of ATP can be synthesised but the levels of free O_2 remain low because oxygen is used up as fast as it is released. Root nodules, in effect, provide nitrogen-fixing bacteria with carbon substrates for respiration and a specialised, low-O_2 environment where nitrogenase can work. The nodules found on non-legume plants do not contain leghaemoglobin, and in that case the mechanisms that achieve low O_2 are still not understood.

In aquatic ecosystems, photosynthetic cyanobacteria are the chief nitrogen fixers and they use a different strategy. The process of photosynthesis generates oxygen, so some filamentous multicellular cyanobacteria carry out the process of nitrogen fixation in specialised cells called **heterocysts** that occur at intervals along the bacterial filament (Figure 2.20).

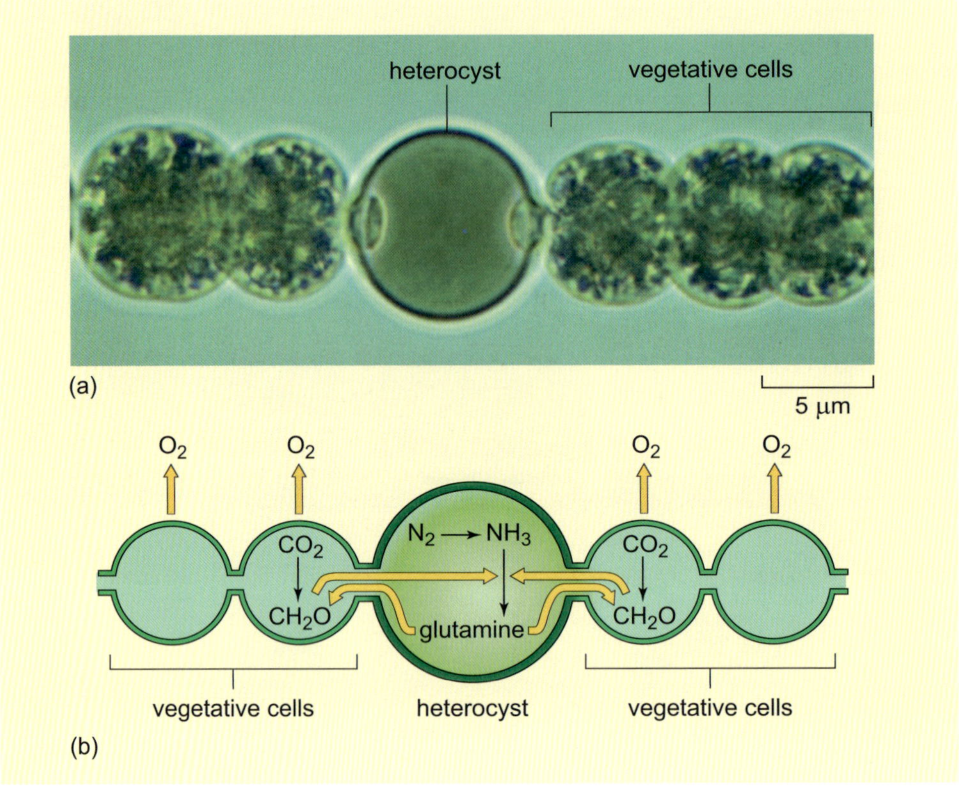

Figure 2.20 (a) Filament of the cyanobacterium *Anabaena* showing the two different cell types: vegetative cells and heterocysts. (b) Nitrogen fixation is restricted to the heterocysts which provide an anaerobic environment for nitrogenase to function. CH_2O in the vegetative cells denotes organic carbon compounds.

The heterocysts have thick walls that are relatively impermeable to O_2. In addition, heterocysts do not generate O_2 during photophosphorylation, even though cyanobacteria use water as their electron donor. They do this by using what is known as cyclic photophosphorylation, which uses photosystem I (PSI) but not photosystem II (PSII) and hence does not generate any O_2 (Book 2, Figure 3.24). In fact, heterocysts do not possess PSII. Cyclic photophosphorylation does not generate any NADPH but does produce ATP. Heterocysts (Figure 2.20) do not fix CO_2, so all the ATP produced from cyclic photophosphorylation is available for nitrogen fixation. Fixed nitrogen is exported from the heterocysts to the neighbouring vegetative cells as the amino acid glutamine. Conversely, organic carbon compounds that are required for biosynthesis are supplied to heterocysts from the neighbouring vegetative cells, which have both PSI and PSII, so there is genuine cooperation between different cell types within filaments of the cyanobacterium.

2.3.4 Endosymbiosis in deep-sea hydrothermal vents

Vast areas of the oceanic sea beds are cold, dark, anaerobic, nutrient-poor and hence relatively lifeless. In 1977 however, deep-sea **hydrothermal vents** emitting plumes of geothermally heated water were discovered along the mid-ocean ridges where the Earth's crust is being formed. The vents support specialised, diverse and highly productive ecosystems (Activity 2.2).

In the vent regions, seawater penetrates through cracks in the sea floor, where it is heated by the molten magma that lies beneath the Earth's crust, to around 400 °C. It re-emerges as hot hydrothermal fluid that is acidic, anaerobic and enriched with various chemicals, including heavy metals, methane, and hydrogen sulfide (H_2S). The hot fluid mixes with the cold (2–3 °C) oxygenated deep-sea water to form warm oxygenated fluid with temperatures in the range 10–30 °C. The mixing of hot hydrothermal fluids and cold seawater can cause the precipitation of minerals dissolved in the vent fluid, forming cylindrical chimney-like structures known as 'black smokers', which may reach 10–15 m in height. Black smokers emit hydrothermal fluids at temperatures of 350–400 °C and at relatively high speeds, which reduces the mixing with cold, oxygenated water.

Figure 2.21 Giant tube worms from deep-sea hydrothermal vents.

Hydrothermal vents support thriving communities of invertebrates: giant tube worms over 2 m long (Figure 2.21) and large numbers of shrimps, giant clams and mussels. The intriguing question is: how do these animal communities exist in the absence of photosynthetic organisms as primary producers?

The 'black smokers' and the warm hydrothermal fluid nearby are inhabited by rich microbial populations.

- Considering the physical and chemical conditions of the warm (10–30 °C) hydrothermal fluid surrounding the vents, suggest what types of microbe would be absent.

□ (i) Light does not penetrate this deep in the oceans, so phototrophs would be absent. (ii) Methanogens would not be expected, because the waters are oxygenated; these archaea are obligate anaerobes.

- What three types of microbe might be present in the warm hydrothermal fluid?

□ (i) Acidophiles; (ii) aerobes; (iii) chemoautotrophs.

The warm water is rich in hydrogen sulfide and oxygen, and supports chemoautotrophic sulfur bacteria, which obtain energy by oxidation of sulfur compounds (Book 2, Section 3.7.2) and use the energy to fix carbon dioxide via the Calvin cycle (Book 2, Section 3.8.4). The energy-yielding reaction is of the form:

$$H_2S + 2O_2 \rightarrow SO_4^{2-} + 2H^+$$

The sulfur bacteria are the primary producers in this ecosystem, and they supply organic carbon to a variety of heterotrophs. Thick mats of sulfur bacteria such as *Beggiatoa* and *Thiobacillus* are common in the water near hydrothermal vents. Microbes resistant to heavy metals such as copper, zinc and cadmium have also been found. The relatively new science of ecological (or environmental) genomics (the study of genetic material recovered from environmental samples) has revealed the enormous diversity of different species of bacteria living near the sulfide-emitting hydrothermal vents. There are about 10 times more species of bacteria than archaea.

■ In contrast to the sulfur bacteria of the warm hydrothermal fluid, what type of microbe would you expect to find in the very hot vent fluid of the 'black smokers'?

☐ The high temperatures should favour hyperthermophiles. The vent fluid is also anaerobic because there is less mixing with oxygenated seawater, so anaerobes should also be prevalent.

The temperature of 'black smoker' emissions, at around 350 °C or more, is too high to support even prokaryotic growth. However, hyperthermophiles have been isolated from the slightly cooler zones. All of these microbes are archaea, most are obligate anaerobes, and they include heterotrophs and methanogens. Recall that *Pyrolobus fumarii* with one of the highest known maximum growth temperatures of 113 °C was isolated from the walls of a 'black smoker' (Figures 2.4 and 2.5).

The animals of these deep-sea hydrothermal vent ecosystems – tube worms, bivalves, gastropod molluscs, annelids and shrimps – are all dependent on the primary production of organic carbon by the chemoautotrophic bacteria. Some of these animals live in endosymbiotic associations with bacteria. The most famous example is the giant tube worm *Riftia pachyptila*, which has no digestive tract, but contains large populations of chemoautotrophic bacteria in a specialised tissue known as the trophosome (Figure 2.22), which constitutes about half the animal's volume.

So far, the endosymbiotic prokaryotes have been found to be non-culturable in the laboratory (Box 2.1), although DNA analysis has identified them as bacteria. All appear to be chemoautotrophic, and key enzymes of H_2S oxidation and CO_2 fixation have been identified in the trophosome tissue. There is no doubt that the tube worm derives its entire organic carbon and energy requirements from the chemoautotrophic activity of the bacteria. In return, the animal supplies the bacteria with their inorganic requirements: H_2S, CO_2 and O_2. Some of the CO_2 comes from the worm's respiration.

The interaction between the tube worms and the chemoautotrophic bacteria is clearly an example of mutualistic endosymbiosis. Both partners gain from the interaction, and the structure of the worm is adapted to accommodate the bacteria.

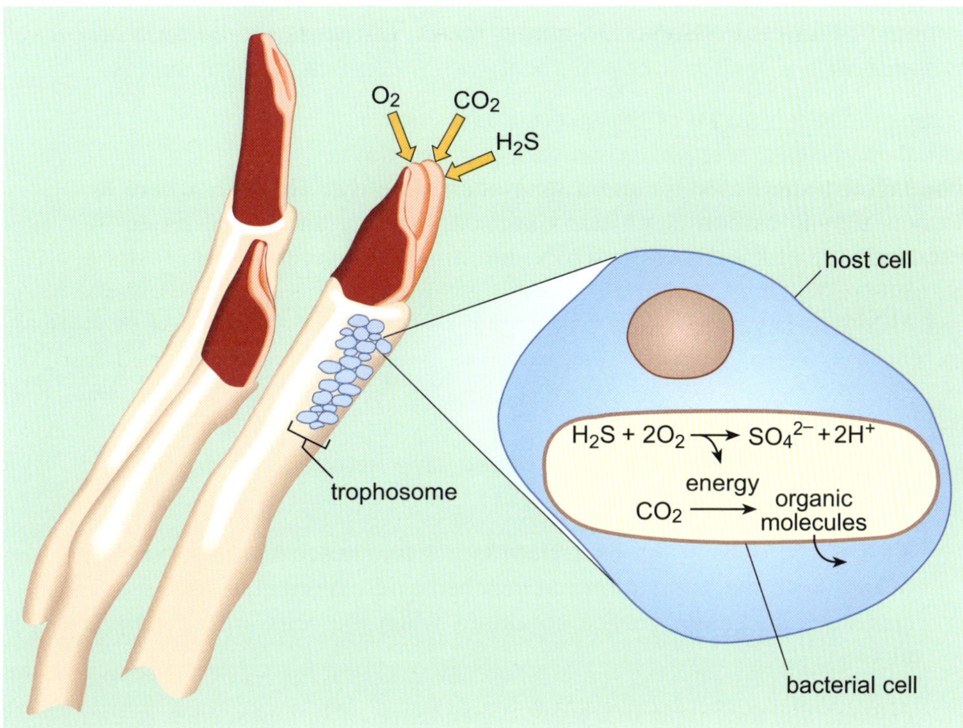

Figure 2.22 The giant tube worms (*Riftia pachyptila*) of hydrothermal vents form endosymbiotic associations with chemoautotrophic bacteria, which are located in a specialised tissue, the trophosome. The worm absorbs O_2, H_2S and CO_2 through its gills and transports these compounds to the bacteria. In return, the bacteria convert these substrates into organic compounds which supply the carbon and energy needs of their worm hosts.

As yet, it is unknown if the bacteria can live outside the endosymbiotic association. It seems likely that both partners are highly adapted both structurally and functionally, such that neither can live alone. The giant clams and some shrimps of the hydrothermal vent communities live in similar endosymbiotic associations.

2.3.5 Mutualism: bioluminescent bacteria

Another interesting group of microbes have the property of emitting light (luminescence). Luminescent organisms, which include certain bacteria, algae, fungi, jellyfish, fish, insects, and squid, have been found in marine, freshwater and terrestrial habitats, but the predominantly marine **bioluminescent bacteria** are the most widespread and abundant. These bacteria belong principally to two Gram-negative genera, *Aliivibrio* and *Photobacterium*, which are usually found associated with fish and squid, some of which possess a special organ where the bacteria grow in mutualistic, endosymbiotic associations (Figure 2.23).

What are the benefits to each of the partners in these mutualistic relationships? For a host fish, light emission by the bacteria provides an advantage for avoiding predation: for example, by frightening or diverting predators. Luminescence from the light organs may also be used for intraspecies

Figure 2.23 The flashlight fish *Photoblepharon palpebratus* has a semicircular light organ beneath the eye, which can be covered by a membrane. The light organ contains the bacterium *Aliivibrio fischeri*.

communication, for example, to attract mates. In return, the bacteria are provided with a sheltered environment with a constant nutrient supply.

Light production involves the enzyme luciferase, which catalyses a reaction in which a reduced flavin mononucleotide ($FMNH_2$) and a long-chain aldehyde (RCHO, where R denotes a hydrocarbon chain) are oxidised, forming an excited flavin intermediate, which decays to FMN, emitting blue–green light:

$$FMNH_2 + O_2 + RCHO \xrightarrow{\text{luciferase}} FMN + RCOOH + H_2O + \text{light}$$

The generation of luminescence has been found to be dependent on bacterial cell density, as shown in Figure 2.24.

■ From Figure 2.24, what is the relationship between cell density and luminescence in *A. fischeri*?

☐ At low cell density the bacterium is not luminescent. Light emission occurs once the population has reached mid-exponential phase (Figure 1.2). Luminescence continues when the cells are in stationary phase.

The explanation for this cell density dependence of luminescence is that expression of the luciferase enzyme is under a form of regulation known as *autoinduction*. The bacteria produce an autoinducer, which accumulates in the local environment as the population density increases. When the concentration of the autoinducer reaches a critical concentration, bioluminescence occurs (Figure 2.24). This mechanism is termed **quorum sensing** because of its density dependence; the term is a reference to boardroom protocol where a quorum, or minimum number of participants, is required for action.

Autoinduction accounts for the fact that free-living bioluminescent bacteria are not luminous. The autoinducer cannot accumulate to high enough concentrations in free-living species. In the light organs of fish (and squid), the bacterial populations reach sufficiently high densities for the autoinducer to accumulate and quorum sensing to occur, which switches on luminescence and the fish light organs emit light.

The autoinducer in *A. fischeri* is *N*-(3-oxohexanoyl)-homoserine lactone (OHHL), a long-chain fatty acid derivative. When living freely in seawater, at cell densities of 10^2 ml^{-1}, the autoinducer diffuses out of the cells, without observable luminescence. In contrast, in the fish light organs, where bacterial cell densities reach 10^{11} ml^{-1}, the autoinducer accumulates and reaches the concentration required for activation of the luminescence (*lux*) genes. The *lux* genes of *A. fischeri* are organised in two operons (Book 1, Section 5.7.1), which are transcribed in opposite directions (Figure 2.25). The *luxA* and *B* genes encode the subunits of the enzyme luciferase, while *luxC*, *D* and *E* encode enzymes involved in synthesis of the aldehyde substrate. In a separate operon *luxI* produces a protein, OHHL synthase, that synthesises the autoinducer and the *luxR* gene encodes an activator protein (luxR) that binds OHHL.

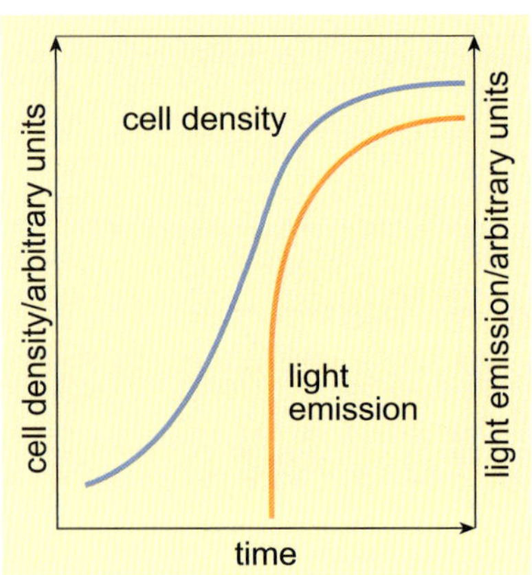

Figure 2.24 Growth and luminescence in cells of *A. fischeri*.

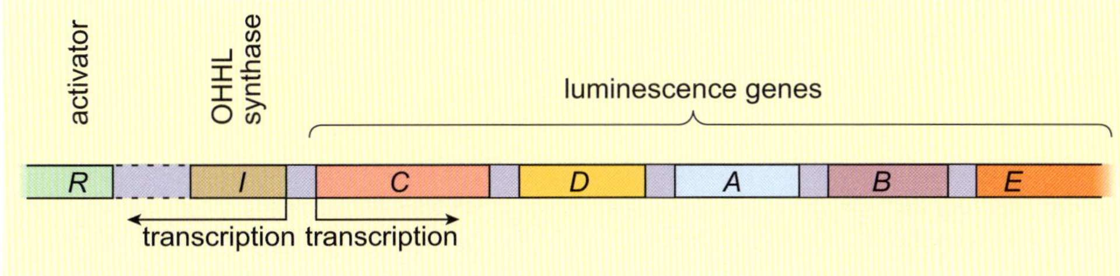

Figure 2.25 The *lux* gene operon in *A. fischeri*.

Figure 2.26 shows how quorum sensing operates at the molecular level when bacterial cells are in the fish light organ. Once the concentration of the autoinducer OHHL reaches a critical threshold, it enters a neighbouring cell. There it binds to the inactive protein, luxR, so forming active luxR, which activates the *luxCDABE* gene cluster, increasing transcription by approximately 1000-fold.

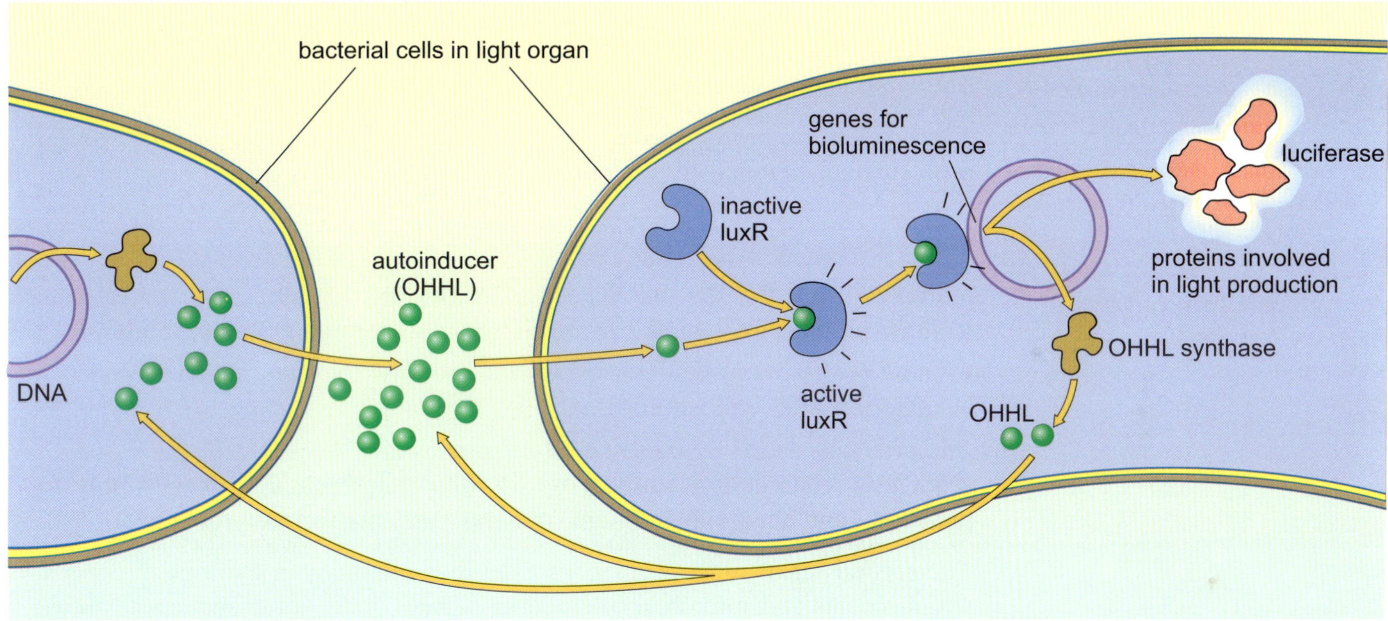

Figure 2.26 Autoinduction of bioluminescence in cells of *A. fischeri* in the light organ of the fish *P. palpebratus*. The autoinducer (OHHL) released from one bacterial cell enters another cell nearby and binds to the inactive luxR protein, so forming active luxR. This switches on the genes for bioluminescence and other proteins are synthesised: the enzyme luciferase produces light; OHHL synthase produces more autoinducer.

- What effect does this increase in transcription have on the level of the autoinducer OHHL?

□ When the *lux* gene cluster is activated, OHHL synthase is produced, which synthesises OHHL; more autoinducer is therefore produced.

Quorum sensing therefore involves a signalling cascade (Book 2, Section 4.4). At first, a few cells send out a message (the autoinducer, OHHL), and the signal spreads throughout the bacterial population in the fish light organ, thus amplifying the response. Regulation of luminescence by quorum sensing ensures that the bacteria waste little energy on light production until effective cell densities have been reached.

Quorum sensing also has a more general role in bacterial communication. It is, for example, involved in infection and the development of disease. The pathogenic bacterium *Pseudomonas aeruginosa* can grow within an animal host without harming it and without alerting its immune system, until the bacteria reach a threshold concentration when they switch on the expression of virulence genes including those encoding toxins which cause disease symptoms. It is hoped that treatments for *Pseudomonas aeruginosa* infections can be developed that work by disrupting quorum sensing (discussed later in Section 4.2.8).

To complete this section, Activity 2.2 provides some video resources that illustrate some of the interactions between microbes and other organisms described here.

Activity 2.2 Microbes in the environment

(LOs 2.1 to 2.4) Allow 1 hour

There are eight video sequences in this activity.

The first set, on protist feeding and predation, examines some of the special problems encountered by unicellular organisms, notably protist autotrophs (phytoplankton) and heterotrophs (flagellates and ciliates) and some of their interactions. The videos show:

- examples of protists moving and feeding
- how protists are sampled from the sea and from deep underground
- how protists may be detected *in situ* and how they are observed and detected after collection.

The problems faced by these small organisms and discussed in the video sequences relate mainly to the viscosity of water, fluctuating food supply for predators, and access to light and mineral nutrients for autotrophs.

The second set of video sequences examines microbes in extreme environments and microbial interactions. These short video sequences show:

- sampling microbes from an extreme environment, namely ice
- extremophiles in the Great Salt Lake
- interactions in the rumen
- the interaction between bioluminescent bacteria and squid
- deep-sea hydrothermal vents and some of the organisms that inhabit these extreme environments.

Summary of Section 2.3

- Microbes interact with other organisms in a wide variety of ways. Interactions occur with other microbes, plants and animals. The symbiotic interactions can be classified as mutualism (including syntrophy or cross-feeding), commensalism, endosymbiosis, parasitism, allelopathy or predation.

- *Bdellovibrio* are bacteria that prey on and invade other prokaryotes.

- The rumen is a specialised stomach in certain mammalian herbivores where the decomposition of complex plant material takes place by the action of anaerobic microbes. The rumen illustrates mutualism, syntrophy and competition. Complex polymers such as cellulose are first degraded to monomers, which are then fermented to short-chain fatty acids (which can be used by the ruminant animal for respiration), plus gaseous CO_2 and H_2. Finally, methanogens use these gases to produce methane, in which syntrophy, involving interspecies H_2 transfer, plays a key role.

- Nitrogen fixation occurs in a small number of species of bacteria and archaea, through the activity of the nitrogenase enzyme complex. Nitrogenase is strongly inhibited by oxygen and operates only in low-oxygen environments. Mechanisms for removing or excluding O_2 include: (a) endosymbiotic associations with plants in which rhizobial bacteria live in root nodules that provide a low-oxygen environment through the action of O_2-binding proteins (leghaemoglobins); (b) specialised low-oxygen cells (heterocysts) in multicellular cyanobacteria.

- Deep-sea hydrothermal vents consist of warm or very hot water rich in inorganic nutrients, notably H_2S, which supports a specialised chemoautotrophically-based ecosystem. Many novel sulfur bacteria and hyperthermophiles have been isolated from these vents. Endosymbiotic mutualism occurs between vent animals and sulfur bacteria.

- Bioluminescent bacteria emit light by the action of luciferase, which is autoinduced when the bacterial populations reach sufficiently high density and hence the autoinducer is at a critical concentration, which activates luminescence genes. This mechanism is termed quorum sensing. These bacteria usually grow in endosymbiotic, mutualistic association in light organs of certain fish and squid.

2.4 Final word

This chapter has considered microbial cells in different natural habitats with a focus on some adaptations that allow them to inhabit a diverse range of environments, particularly extreme environments, notably temperatures above the boiling temperature of water, pH values near zero, incredibly high pressures at the bottom of oceans and very high levels of ionising radiation that would disrupt the human genome speedily and irreversibly!

It should be clear to you that many microbes do not live a lone existence but interact with a huge range of other organisms, both other microbes and also macroscopic plants and animals. Many such associations are mutually beneficial, so that both organisms in the partnership benefit from the interaction; indeed, without such associations the individual organisms would not survive. Those interactions in which the microbe alone gains from the association, while the host animal or plant is significantly debilitated or even killed have been mentioned only briefly. Such pathogenic interactions often bring about diseased states in the host animal or plant, and this is the first subject of the next chapter.

2.5 Learning outcomes

2.1 Distinguish between different types of microbes and the major categories of extremophiles.

2.2 Discuss some of the problems of studying microbial cells in their natural environments.

2.3 Describe cellular and molecular adaptations to challenging environmental extremes, including low or high temperature and pH, oxygen availability, high salt and high pressure.

2.4 Discuss some ways in which microbes interact with other microbes or eukaryotes, using named examples of mutualistic, endosymbiotic and predatory associations.

References

Bartlett, D.H. (2002) 'Pressure effects on in vivo microbial processes', *Biochimica et Biophysica Acta*, vol. 1595, pp. 367–81.

Cox, M.C. and Battista, J.R. (2005) '*Deinococcus radiodurans* – the consummate survivor', *Nature Reviews Microbiology*, vol. 3, pp. 882–92.

Dalhus, B., Saarinen, M., Sauer, U.H., Eklund, P., Johansson, K., Karlsson, A., Ramaswamy, S., Bjørk, A., Symstad, B., Naterstad, K., Sirevåg, R. and Eklund, H. (2002) 'Structural basis for thermophilic protein stability: structures of thermophilic and mesophilic malate dehydrogenase', *Journal of Molecular Biology*, vol. 319, pp. 707–21.

Hewitt, J. and Morris, J.G. (1975) 'Superoxide dismutase in some obligately anaerobic bacteria', *FEBS Letters*, vol. 50, pp. 315–18.

Latham, M.J. and Wolin, M.J. (1977) 'Fermentation of cellulose by *Ruminococcus flavefaciens* in the presence and absence of *Methanobacterium ruminantium*', *Applied and Environmental Microbiology*, vol. 34, pp. 297–301.

Nichols, D.S., Miller, M.R., Davies, N.W., Goodchild, A., Raftery, M. and Cavicchioli, R. (2004) 'Cold adaptation in the Antarctic archaeon *Methanococcoides burtonii* involves membrane lipid unsaturation', *Journal of Bacteriology*, vol. 186, pp. 8508–15.

Trampuz, A., Piper, K.E., Steckelberg, J.M. and Patel, R. (2006) 'Effect of gamma irradiation on viability and DNA of *Staphylococcus epidermidis* and *Escherichia coli*', *Journal of Medical Microbiology*, vol. 55, pp. 1271–5.

Chapter 3 Cells and disease

3.1 Introduction

Up to this point in the module you have concentrated primarily on the processes that occur in normally functioning cells. In this chapter you will focus on a variety of abnormal cell behaviours and investigate some of the factors that can bring about changes in cell function leading to disease.

Conventionally, the word 'disease' is used to encompass any type of malfunction in a part or parts of a living organism. The variety of underlying causes of disease is truly staggering and the size and complexity of this topic means that it is not possible, in this one chapter, to discuss all of the cellular aspects of disease. Instead, you will focus on four case studies drawn from three major categories of disease: infections, inherited conditions and the degeneration of bodily systems which frequently occurs as we age.

The four case studies look at bacterial infections, exemplified by MRSA; viral infections, using foot-and-mouth disease as an example; inherited diseases, namely xeroderma pigmentosum, Huntington's disease and Marfan syndrome as examples; and finally, degenerative disorders, focusing on Parkinson's disease. Studying these examples will allow you to fully appreciate the wide-ranging nature of different disease states: from infectious diseases that originate from outside the organism to inherited diseases which result from the transmission of faulty genes, with a predictable pattern of inheritance, from one generation to the next. Studying degenerative diseases will highlight their complex causes; allowing you to appreciate that, although there is some evidence to indicate they have a genetic component, the influence of the environment on the individual is also highly significant.

Each case study will be introduced by a discussion of the underlying cause of the disease at a cellular level and the resulting disease symptoms. Then you will be directed to the associated Activities (3.2–3.5) to study supplementary information where you will explore the impact of the disease on individuals and communities.

Of all these types of disease, only bacterial pathogens bring about cellular damage from the outside; they act at the extracellular level. The other types of disease you will investigate act within cells: either because an infectious agent penetrates the cell (as in the case of viruses) or because there is an inherent fault within the cell which means it cannot carry out its normal functions (for example, inherited disease or degenerative disease). You will begin this exploration of disease processes by looking at bacterial infections and the defence mechanisms that exist to limit the cellular damage they may cause.

3.2 Bacteria: disease through damage

Although a few bacteria such as *Mycobacterium tuberculosis* can penetrate and infect living cells, most diseases caused by bacteria are all about damage caused at the level of tissues – where bacteria growing and multiplying amongst the host's living cells wreak havoc.

You are already familiar with the Bacteria as prokaryotic cells (Book 1, Chapters 2 and 3).

- What features of the Bacteria are distinctive when compared to eukaryotic cells?

- Prokaryotic cells have no nucleus, no membrane-bound organelles and they possess very different surface characteristics to eukaryotic cells. In addition to a cell membrane, the bacterial cell is surrounded by a cell wall. In the Bacteria this is made of the amino sugar, peptidoglycan, a polymer composed of sugars linked by amino acids. Recall from Book 1, Section 3.3 that the cell walls of Gram-positive bacteria and Gram-negative bacteria differ with respect to the arrangement and thickness of the peptidoglycan layer.

The tissues of eukaryotes provide an appropriate environment in which many types of bacteria can grow and multiply. There is a plentiful supply of nutrients, and the ambient temperature, pH and osmotic balance are generally favourable for many mesophilic bacterial species. Humans are host to several hundred different kinds of microbes, mainly bacteria. The collection of microbial species living in and on a healthy organism is referred to as the **normal flora**, a few of which are identified in Figure 3.1.

To establish themselves within the human body, bacteria must first attach to or penetrate a surface of the host and then multiply in or on the host's tissues. To survive and reproduce, bacteria then have to acquire nutrients from the host and at the same time evade the host's defence mechanisms. For the cycle of infection to continue, the bacteria must multiply efficiently and eventually leave the host's body in sufficient numbers to ensure transmission to a new host.

It must be stressed here that causing disease in the host is usually of no direct advantage to the pathogen; disease is simply a consequence of the proliferation of the pathogen within the host. What is it that makes a particular bacterium pathogenic to its host? It is difficult to generalise, but the symptoms of infection can be broadly classified as (1) those produced directly by the invading bacterium or its secretions and (2) those that are the result of the host's immune system's response to the infection (e.g. fever). In most cases, the disease symptoms, observed at the level of the whole organism, can be explained in terms of the effects of the pathogen at the cellular or molecular level.

3.2.1 Defence against bacterial disease: innate immune defence

Microbial pathogens (not just bacteria and viruses but also fungi and protists) and their hosts have been coevolving for hundreds of millions of years, and this association has led to the development of a variety of host mechanisms for recognising the pathogen, limiting its growth and preventing its spread. Most organisms have some sort of system for combating bacterial infection, ranging from the production of antimicrobial enzymes to phagocytic cells that 'eat' infecting bacteria.

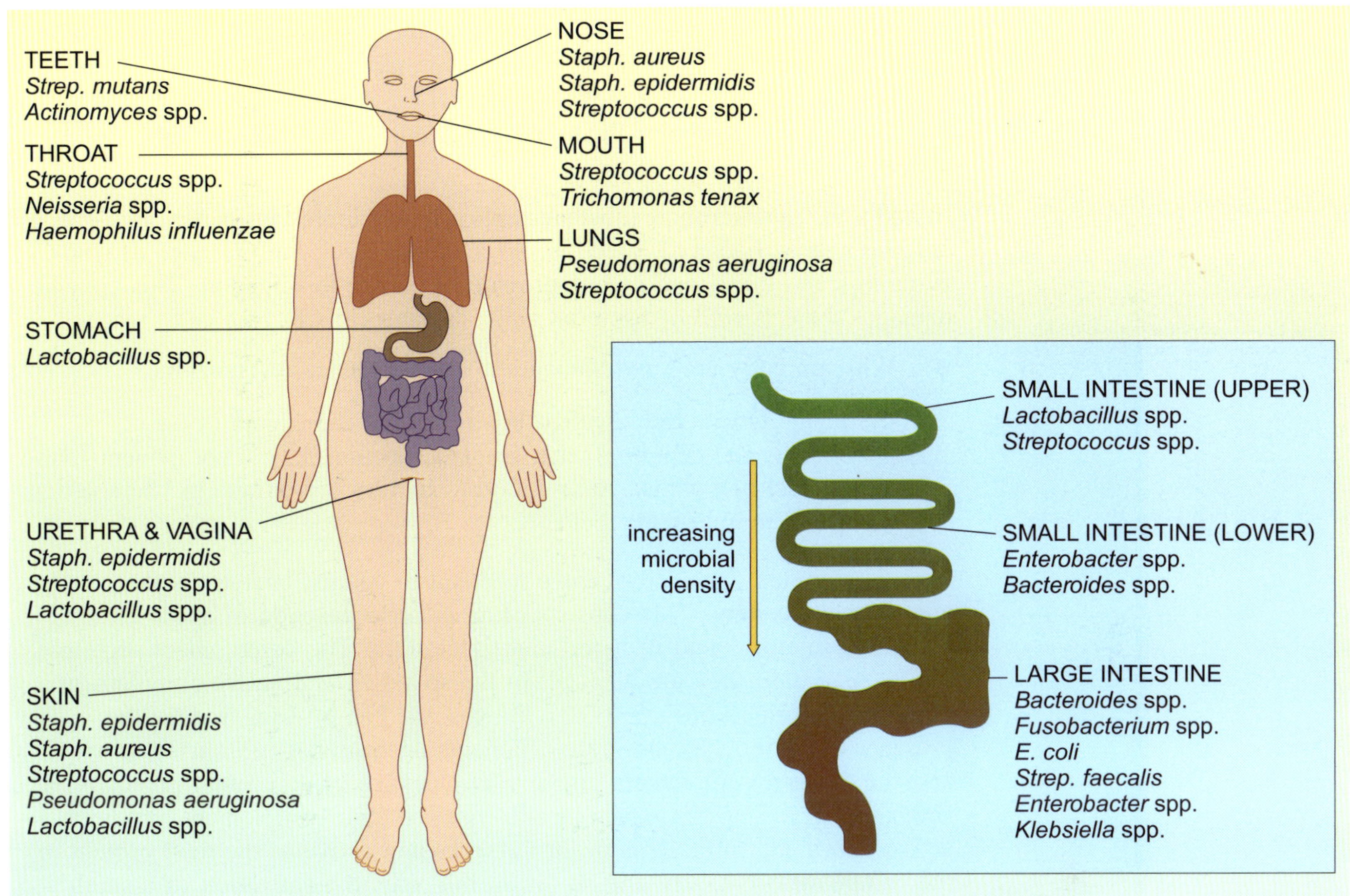

Figure 3.1 Some of the bacteria that make up the normal flora of the human body and their locations in the human body (*Strep.* = *Streptococcus*; *Staph.* = *Staphylococcus*; note that 'spp.' means multiple species.) Characteristics of the microbial flora in different regions of the intestinal tract are shown separately on the right of the figure. It is estimated that of the order of 10^{14} bacteria inhabit the human body, the majority of them in the gut, with microbial density increasing by a factor of more than 10^7 between the stomach and the large intestine.

Vertebrate animals, including humans, have developed the most sophisticated set of defences. The collective name for these mechanisms is the immune system, the protective structures and processes of which can be broadly divided into two defensive systems: the **innate immune system** and the **adaptive immune system**. The innate immune system is unspecific; it provides generalised defence against all invading pathogens. In contrast, the **adaptive system** recognises and targets specific pathogens and develops immunological memory, conferring long-term protection against that pathogen.

The innate immune system provides the first line of defence against most bacterial infections: preventing entry and limiting the growth of pathogens in the body. Firstly, a range of measures helps keep infectious agents out of the body, from the skin's impenetrable barrier and the acidic nature of sweat to the acid in our stomach.

Potential entry points for pathogens, however, are the mucosal surfaces: where invading organisms can come into direct contact with epithelial cells. The respiratory tract, the surface of the eye and the gastrointestinal tract are all vulnerable to attack and yet, they too, have mechanisms that protect them from infection; particularly advantageous is the constant flow of fluid over these surfaces which helps flush pathogens out of the body and prevent attachment.

In addition, antimicrobial substances play a very important part in non-specific defence against bacterial infection. At a simple level, the presence of lysozyme in tears, saliva and mucus provides a barrier to infection. This enzyme damages the cell wall of bacterial cells, causing cells to lyse and then die.

- Why are body cells not damaged by the presence of lysozyme?

- Lysozyme targets the covalent bonds linking the sugars in the polysaccharide, peptidoglycan (Book 1, Section 3.3), a component found in bacterial cell walls – these polysaccharides are unique to bacteria and not found in eukaryotic cells.

A more complex chemical defence mechanism against bacteria is the **complement system**. This consists of an enzyme **cascade** of about 20 different proteins found in blood and tissue fluid. The complement proteins are inactive until they are cleaved. In the so-called alternative complement pathway (described in Figure 3.2), the cascade can be triggered by the presence of lipopolysaccharide components on bacterial cell surfaces, which activate the first complement enzyme by cleavage of its protein precursor. The activated enzyme then cleaves its substrate, another complement protein to its active enzymatic form. This in turn cleaves and activates the next protein in the complement pathway and so on. In this way, the activation of a small number of complement proteins at the start of the pathway is rapidly amplified by activating larger numbers of proteases at each successive enzymatic reaction, resulting in the rapid generation of disproportionately large numbers of complement response molecules. The final step of this cascade produces a **membrane attack complex** which damages the bacterial cell by punching holes in its surface. Once breached in this way, the cell rapidly takes up water by osmosis, causing it to undergo osmotic lysis (i.e. swelling followed by bursting).

Note that there are several complement activation pathways; in contrast to the alternative complement pathway, the classical pathway is activated by the presence of antigen–antibody complexes (Book 1, Section 2.2.3). Antibodies are produced by the adaptive immune system (see Section 3.3.3 in this chapter): they attach to surface antigens on pathogens, marking them out for destruction by cells of the innate immune system (as detailed below). You will return to the role of the adaptive immune system later in this chapter when you will study how organisms respond to viral infections.

There are other innate defence mechanisms that are particularly effective against bacteria. These include so-called **cell-mediated** defence mechanisms: phagocytosis and cytotoxicity. The cells involved in this defence system are white blood cells or **leukocytes**, which act in three distinct ways to mount an immune defence:

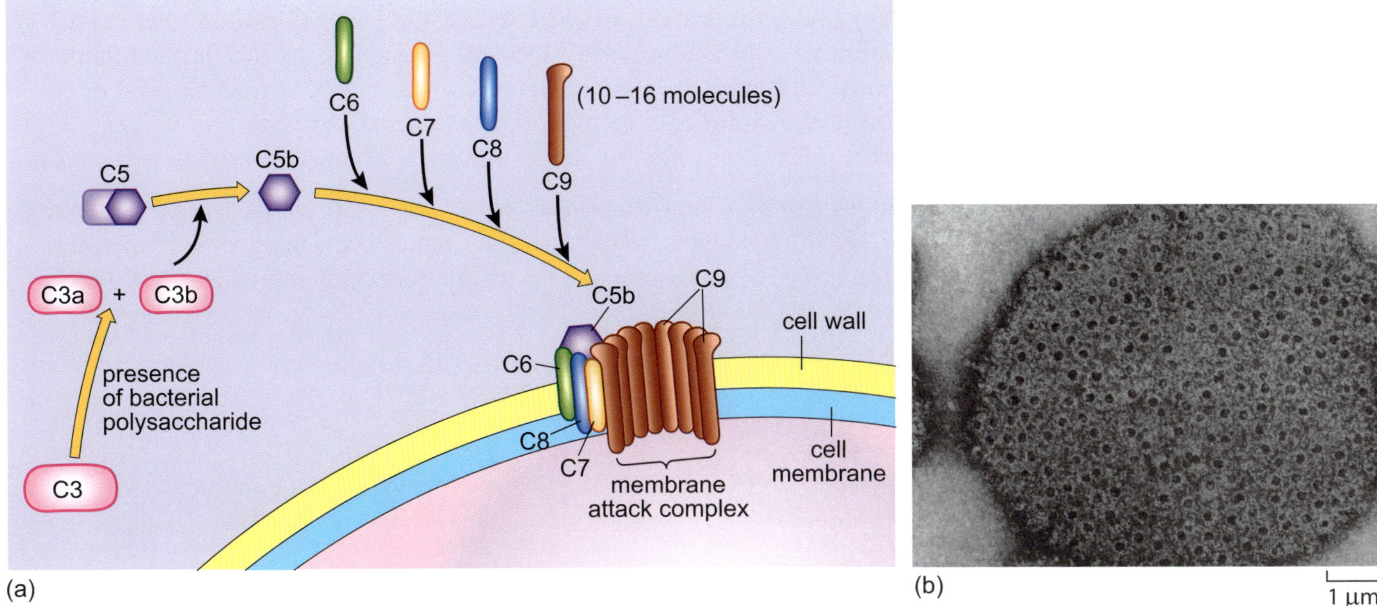

Figure 3.2 (a) Diagram of the alternative pathway for complement activation. This is an innate, non-specific defence mechanism that responds directly to the presence of polysaccharides on the surface of bacterial pathogens. The pathway is activated when one of the proteins in the complement cascade, C3, binds directly to the surface of a bacterium. This event initiates an enzyme cascade leading to the production of the membrane attack complex. This is constructed from molecules of C5b, C6, C7, C8 and several molecules of C9 forming a tube-like structure inserted through the bacterial cell wall and into the cell membrane. Water may now enter the cell by osmosis and cause cell death by osmotic lysis. (b) This shows the cell surface of *Shigella dysenteriae* with numerous holes punched in the cell membrane by membrane attack complexes.

- phagocytosis, i.e. engulfing and digesting potentially harmful material in the body
- cytotoxicity, i.e. killing infectious organisms by damaging the cell membrane of a pathogen or an infected cell
- generating inflammation around the site of an infection.

Several types of leukocytes are phagocytic (Figure 3.3a) and are found throughout the tissues and in the blood. The first step in phagocytosis is attachment to the invading bacteria. Attachment is facilitated by the presence of receptors on the surface of the phagocyte; the binding sites of these receptors recognise the unique molecules found on the surface of bacteria. In addition, some of the complement system proteins may remain attached to the cell wall of bacteria and 'mark' them out for recognition by the receptors on phagocytes. Once attached, local rearrangement of the cytoplasmic actin filaments by the phagocyte (Book 2, Section 5.4.3) allows the cell's cytoplasm to flow around the bacteria and fuse, enclosing the bacterium in a vesicle (Figure 3.3b). Lysosomes fuse with the vesicle (Book 1, Section 3.4.9) and release degradative enzymes, digesting the bacterium into small molecules, which can be absorbed by the cell.

A different class of leukocytes, **cytotoxic cells**, also kill bacterial cells, but they use different methods. The cytotoxic cell lies alongside the bacterium, attaches to its surface and directs a chemical attack on the bacterial cell

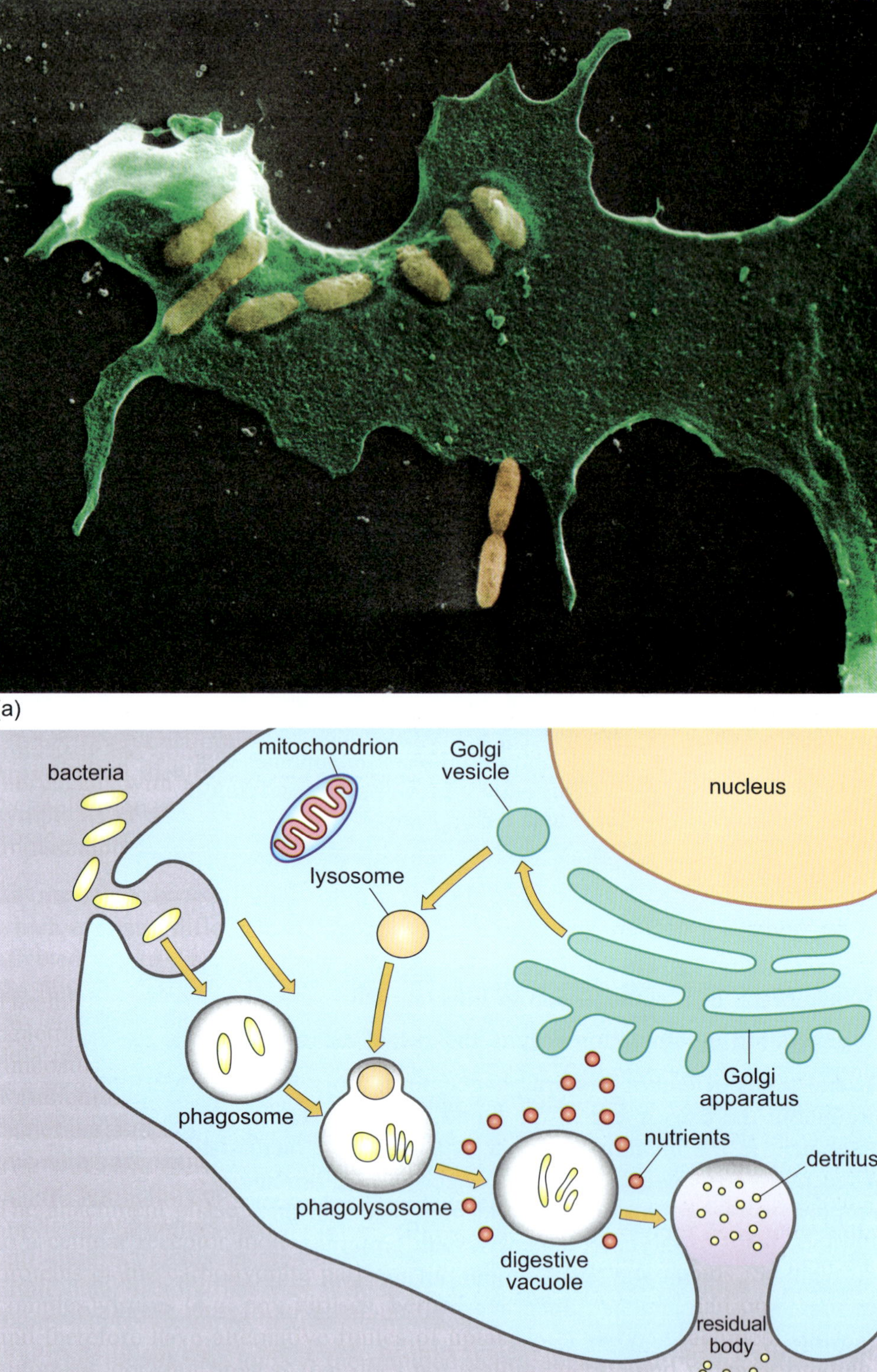

Figure 3.3 (a) Photomicrograph of phagocytosis of bacterial cells by a leukocyte. (b) Diagrammatic representation of the phagocytotic process within the leukocyte.

surface. Depending on the type of cytotoxic cell, this may involve either degradative enzymes or molecules capable of puncturing the bacterial cell membrane. An example of the latter is **perforin**, a cylindrical protein molecule that inserts into the bacterial cell membrane and forms a pore, allowing influx of fluid into the cell, which, in turn, swells up and bursts. This is a similar mechanism to the membrane attack complex generated in the final stage of the complement cascade described above.

The final type of non-specific cell-mediated defence is the inflammatory response. Inflammation tends to occur at the site of infection when invading bacteria have overcome the other non-specific responses described already and the innate immune system needs to ramp up its response to the invading organisms. It is a protective response and, although frequently painful, its purpose is to localise the infection and inactivate and remove the infecting bacteria before too much damage is caused. The presence of bacterial cell wall components and activated complement molecules are primary triggers of inflammation. The external signs of inflammation (redness, swelling and heat) are all due to dilation of blood vessels, increased blood flow to the area and leakage of plasma from the blood vessel into the tissues. The leaking blood plasma brings with it large numbers of leukocytes and complement proteins into the tissues. The fluid dilutes out any bacterial toxins, while the higher concentration of complement proteins increases the level of antibacterial activity, including phagocytosis and cytotoxicity, in the area. Inflammation often results in pus formation or **suppuration**. Pus consists of dead cells, dead phagocytes, debris, fibrin (blood clotting protein), plasma and living and dead bacteria.

3.2.2 Bacterial virulence factors

As host defence mechanisms have evolved and adapted, bacteria have also evolved new characteristics that enhance their ability to evade these mechanisms and to successfully colonise the host i.e. their **virulence**.

Virulence is a measure of the ease with which an organism can cause damage and disease in host tissues: the more virulent the organism, the greater the damage to the host. Even within a single species of bacteria there may be a huge variation in virulence between bacterial strains; some strains show little ability to infect and cause disease, whilst others cause very significant damage. The virulence of a given pathogen is genetically determined by so-called **virulence factors**. These are specific attributes that affect the pathogen's ability to attach to, enter and reproduce inside its host and its capacity to evade the host's immune defences.

Important virulence factors for many bacterial pathogens are exotoxins: toxins that are secreted into the surrounding area. It is often the presence of these substances that damages host cells and brings about disease symptoms. The majority of exotoxins are proteins and some exotoxins are very toxic indeed: for example, just one molecule of the diphtheria exotoxin (secreted by *Corynebacterium diphtheriae*, the pathogenic bacterium that can cause diphtheria) is sufficient to kill a cell and one gram of botulinum toxin (secreted by the *Clostridium botulinum* bacterium that may cause botulism) is sufficient to kill 10 million people.

Exotoxins act in a variety of ways. Some may assist in colonisation of the host (by promoting cell lysis or inhibiting host cell protein synthesis, Table 3.1) or transmission to a new host; but in other cases, such as neurotoxins, exotoxin production has no obvious adaptive value to the pathogen.

Table 3.1 Some bacterial exotoxins of importance in causing major symptoms of disease.

Bacteria	Exotoxin	Tissue/cells damaged	Action	Disease
Clostridium tetani	tetanospasmin	neurons	spastic paralysis	tetanus
Clostridium perfringens	α-toxin (phospholipase C)	red blood cells, white blood cells, endothelial cells	cell lysis	gas gangrene
Clostridium botulinum	neurotoxin	neuromuscular junction	flaccid paralysis	botulism
Staphylococcus aureus	α-toxin	skin and other epithelia	pore formation leading to cell lysis	abscesses, post-operative wound infection
Corynebacterium diphtheriae	diphtheria toxin	throat, heart, peripheral nerves	inhibits protein synthesis	diphtheria
Vibrio cholerae	enterotoxin*	intestinal epithelium	fluid loss	cholera
Bordetella pertussis	pertussis toxin	respiratory epithelium	disrupts signal transduction	whooping cough
Listeria monocytogenes	haemolysin (phospholipase C)	white blood cells	cell lysis	listeriosis
E. coli 0157:H7	verotoxin	intestinal epithelium	severe diarrhoea and kidney damage	gastroenteritis

* An enterotoxin is a microbial exotoxin that is present in the gut.

Figure 3.4 shows virulence mechanisms of some different types of exotoxins. Some exert their effects extracellularly: for example, breaking down extracellular material such as collagen or digesting cell membrane phospholipids and destroying cells (Figure 3.4a). This tissue breakdown facilitates the spread of the infection. Cell membranes can also be damaged by the insertion of pore-forming exotoxins, which kills cells by 'punching' holes (pores) in the membrane, enabling water and ions to flow in and out freely and so leading to osmotic lysis (Figure 3.4b).

Other exotoxins enter cells and interfere with a specific metabolic process. These molecules are usually composed of two subunits, with different functions. The 'binding' (B) subunit binds to a membrane receptor on a host cell and as a result of binding, the active (A) subunit is taken into the cell where it exerts its toxic effect. Diphtheria and cholera toxins are of this type. Both exert their effect by covalently modifying key intracellular proteins. Diphtheria toxin blocks protein synthesis, by inactivating a protein factor required for efficient polypeptide elongation (Book 1, Section 6.6.3), resulting in cell death (Figure 3.4c).

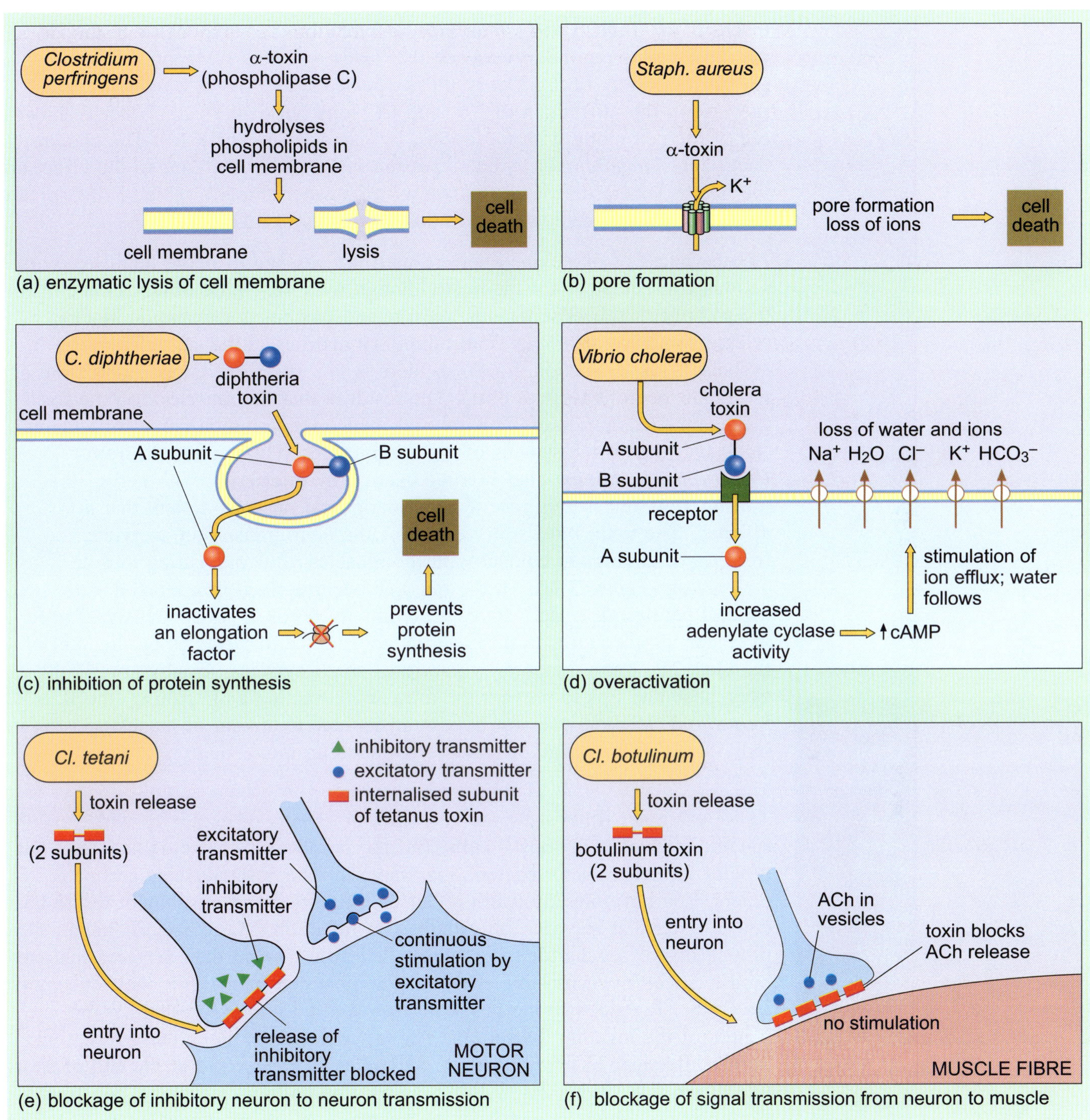

Figure 3.4 The mechanism of action of some exotoxins. (a) Cell membrane lysis. (b) Formation of pores in the cell membrane. (c) Inhibition of protein synthesis. (d) Overactivation of cAMP-regulated ion efflux. (e) and (f) Blockage of neurotransmitter release (ACh = acetylcholine). The result of (e) is spastic paralysis (muscle spasms), whereas (f) leads to flaccid paralysis.

All eukaryotic ribosomes are sensitive to diphtheria toxin, in fact sensitivity is a characteristic used in prokaryote taxonomy to distinguish Archaea (whose ribosomes are sensitive to the toxin) from Bacteria (whose ribosomes are insensitive). Cholera toxin interferes with one of the components of the G protein-coupled regulatory cascade that leads to cyclic AMP production (Book 2, Section 4.6.1). The result is a raised cAMP level which causes an increase in the movement of Na^+ and Cl^- ions out of the cell. Water is drawn out of the cell along with these ions, by osmosis (Figure 3.4d). The cholera bacterium, *Vibrio cholerae*, colonises the intestinal epithelium; so the effect on the host is severe diarrhoea (known as 'rice-water stools'), which helps spread the infection and leads to dehydration, which can be fatal.

Another class of exotoxins interfere with the passage of nerve impulses, with serious consequences for the host although with very little obvious benefit to the pathogen itself. The tetanus toxin tetanospasmin, for example, blocks release of neurotransmitter from inhibitory neurons in the central nervous system (Book 2, Section 4.2.4), so there is too much excitatory stimulation of the motor neurons (Figure 3.4e). The result is that the muscles become fixed in a contracted state (spastic paralysis); hence the common name for tetanus, 'lockjaw', due to the spasms of the jaw muscles. The neurotoxin from *Clostridium botulinum* also exerts its action by interfering with nerve transmission, but in this case it is the peripheral nervous system that is affected. The toxin blocks the release of the neurotransmitter acetylcholine from nerve endings in contact with the muscles, thus preventing muscle contraction (Figure 3.4f). So the muscles become fixed in a relaxed state (flaccid paralysis). A derivative of botulinum, botox is now extensively used in cosmetic procedures to disguise the appearance of wrinkles on the skin of the face. The toxin temporarily paralyses small muscles, usually around the eyes, nose and lips to prevent the muscle contracting and 'pulling' the skin into a fold. Medical uses for botulinum include treatment of muscle spasms in individuals with a range of conditions including motor neurone disease and cerebral palsy.

These examples illustrate some very specific effects of exotoxins with very specific effects on the host's cells. But the production of exotoxins is just one type of virulence factor bacteria may have in their arsenal; there are many others. You have already learnt about the importance of attachment to the host cell in order that bacteria can establish an infection. Many bacteria have protein or polysaccharide molecules called **adhesins** on their surfaces that can bind to specific receptor molecules on the host cell membrane; these receptors have other functions in the host cell: they are not simply present for the benefit of the invading bacterium. The latter simply exploits these receptors during infection. In many bacteria, the adhesins are located at the end of short pili (Book 1, Sections 2.3, 3.3 and 5.7.5), fine hollow extensions, constructed from protein, which extend from the bacterial cell membrane (Figure 3.5). The adhesin on the tip of each pilus recognises the polysaccharide chains of a specific glycoprotein or glycolipid receptor on the eukaryote cell surface, allowing attachment. The type of adhesin present determines the bacterium's virulence in an organism. Some bacteria are even capable of switching the pilus type they express dependent on the host they are infecting; ensuring there is a good match between the pilus expressed and receptors on the host cell membrane.

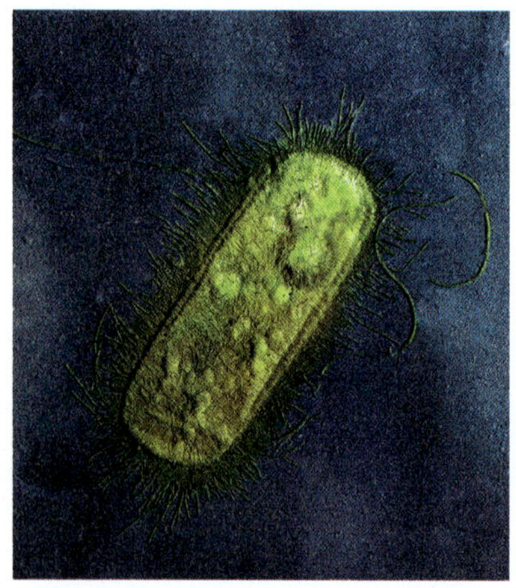

Figure 3.5 Pili on the surface of *E. coli*. Pili are used to attach bacteria to the surfaces of other cells, to environmental surfaces and to each other during conjugation (sex pilus). Note that some sources describe short pili as fimbriae. The longer projections (particular visible on the right-hand side) are flagella.

Not all strains of a particular bacterial species possess the same virulence factors. A good example of a bacterial species that demonstrates variable virulence is *E. coli*. Most strains of *E. coli*, despite being present within the human gastrointestinal tract in very large numbers, are completely harmless to their host. Unfortunately, several strains of pathogenic *E. coli* exist: so-called **enteropathogenic** strains, which cause very severe diarrhoeal disease, particularly in infants, throughout the world. The difference in virulence between these strains and the *E. coli* of our normal gut flora is the enhanced ability of the pathogenic *E. coli* to bind to the intestinal mucosa. All *E. coli* are capable of tethering themselves to the intestinal mucosa but the virulent strains possess pili that are particularly efficient at intimately attaching to the epithelial lining of the small intestine (Book 1, Section 2.4.5), an area of the gut not normally colonised by *E. coli*. At least 20 different types of *E. coli* pili have been identified, and enteropathogenic *E. coli* possess a highly distinctive type, called a **bundle-forming pilus**. These pili firmly tether the bacterium to the wall of the intestine despite the constant churning of the gut contents, allowing the bacteria to colonise the host gut epithelium and cause enteric disease. After initial attachment, the presence of the bacteria 'effaces' or removes the microvilli on the surface of the epithelial cells, forming a lesion.

- What effect might the loss of microvilli on the surface of the intestinal epithelial cells have on the functioning of the gut?

- The microvilli increase the surface area for absorption of nutrients and water from the gut (Book 2, Section 2.4.5), so the loss of microvilli will reduce the efficiency of absorption of these substances across the gut wall.

Once the microvilli are lost from the immediate gut wall, a complex sequence of events occurs which binds the invading bacterium even more intimately to the epithelial cell in the area of the lesion. Activity 3.1 will allow you to study, in detail, the complex mechanism this strain of *E. coli* uses to bind to the gut lining.

Activity 3.1   Enteropathogenic *E. coli*: mode of attachment

(LOs 3.1, 3.2 and 3.4) Allow 20 minutes to view the animation and read the associated notes

In this activity you will watch an animation which explores the process by which enteropathogenic *E. coli* attaches itself to the cells of the human gut via a two-stage process; first via pili which recognise and bind to surface molecules on the epithelial cells lining the gut and second via insertion of its own receptor molecules (Tir) into the epithelial cell.

Some notes, diagrams and questions, to aid your understanding, accompany the animation.

Like enteropathogenic *E. coli*, other bacterial species also produce a very wide array of virulence factors which attack the cell on all fronts – this diverse approach is common in some of the most successful pathogens and allows them to colonise their host very rapidly, bringing about whole-scale destruction of host tissues. In the next section you will investigate the multiple virulence factors of one strain of *Staphylococcus aureus*; the so-called superbug MRSA (methicillin-resistant *Staphylococcus aureus*) and the infections it can cause.

3.2.3 Case study 1: Bacterial disease: MRSA – a superbug?

S. aureus is a Gram-positive spherical bacterium (Figure 3.6) which naturally colonises the nasal passages and skin. Some 30% of the human population carry the bacteria persistently and yet are asymptomatic (i.e. have no physical symptoms of disease).

■ The relationship between *S. aureus* and its asymptomatic human host is described as commensal. Explain what this term means in this context.

☐ Commensalism describes two species living in close proximity where one organism benefits from the association and the other neither suffers any harm nor derives any benefit from the association (Section 2.3).

Research has shown that those individuals who carry the bacteria asymptomatically are susceptible to future infection and it is thought that these carriers are likely to be the source of infection for others within the population. Carrier individuals can 'infect' themselves, if the colonising bacteria from the nose for example are transferred to another part of the body, or they can transmit *S. aureus* by direct physical contact with other people or by contaminating objects that will be used by other people.

Although common, *S. aureus* is responsible for a range of infections – from the superficial such as boils (Figure 3.7) and styes (an infection of the eyelid), to much more serious infections such as meningitis, pneumonia, food poisoning and toxic shock syndrome. Many of these infections stimulate the formation of pus by the host and are termed pyogenic. This poses a fascinating question: why does *S. aureus* show this surprising range of pathogenicity?

The variation in the pathogenicity of this organism results from the expression of a whole range of different bacterial genes. Recent studies report that within the bacterial genome of *S. aureus*, there are a grand total of 226 genes controlling the expression of different virulence factors. Some of these genes are carried on mobile genetic elements such as plasmids or transposons (Book 1, Section 5.8.2) that can be easily transferred horizontally between different *S. aureus* cells, or even transferred to other bacterial species.

S. aureus virulence factors range from the ability to adhere to solid substrates, creating biofilms, to the production of potent exotoxins like α-toxin (Figure 3.4b) and the acquisition of resistance to antibiotic drugs developed to kill them (Table 3.2). In each case the virulence factor increases the ease with which the organism can inflict damage to the host tissues. In the next two sections you will look at two of these *S. aureus* virulence factors and explore the underlying mechanisms of virulence.

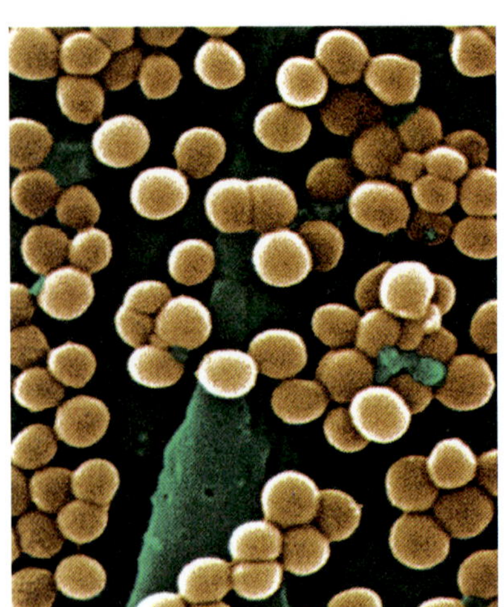

Figure 3.6 Electron micrograph showing methicillin resistant *Staphylococcus aureus* (MRSA).

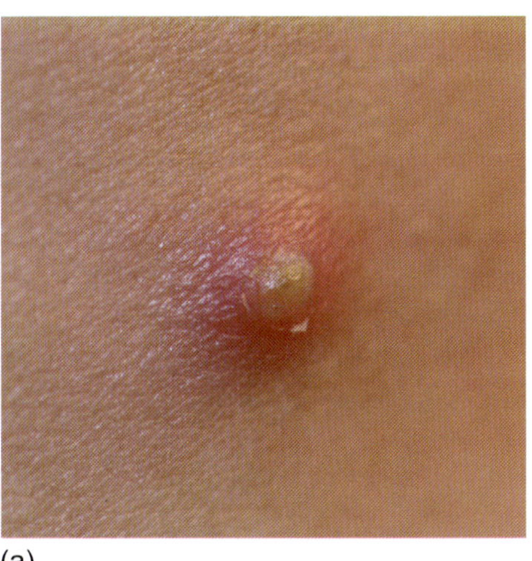
(a)

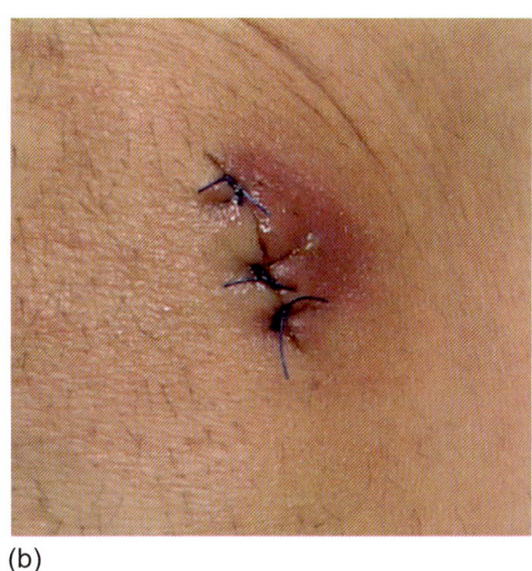
(b)

Figure 3.7 Staphylococcal infections: (a) a boil; (b) an infected wound (the black line shows the three stitches used to close a cut in the skin).

Table 3.2 Some *Staphylococcus aureus* virulence factors.

Type of virulence factors	Mechanism of action	Associated clinical symptoms
Involved in attachment	Microbial surface components which can bind to host molecules such as fibrin and collagen	Binds to bone and joints, leading to osteomyelitis/arthritis. Also can attach to prosthetic devices in the body/catheters
Involved in persistence	Resistance to antibiotics Biofilm accumulation	Inability to treat infections with antibiotics Persists on prosthetic devices in the body/catheters
Involved in evading/destroying host defences	Production of proteins that interfere with antibody binding, and inhibit complement activation Formation of a protective capsule that allows survival inside phagocytes	Invasive skin infections
Involved in tissue invasion/penetration	Production of enzymes, including proteases, nucleases and lipases that facilitate invasion and spread of bacteria through tissues	Tissue destruction
Involved in toxin-mediated disease	Production of enterotoxins (that act on the cells of the gut) Toxins, A and B α-toxin	Vomiting/diarrhoea associated with food poisoning Exfoliation of skin Abscess formation/wound infection

Antibiotic resistance

A feature of *S. aureus* is the ease with which it acquires resistance to antibiotics. **Antibiotics** are antimicrobial compounds, naturally produced by some microbes, that kill or inhibit the growth of other microbes. Antibiotics used to combat bacterial disease are more correctly called antibacterials but the terms antibiotic and antibacterial are often used synonymously. Antibacterials are selectively toxic; that is, they target processes that are unique to prokaryotic cells leaving eukaryotic cells unaffected. The first universally recognised antibiotic was penicillin. This antibiotic targets the production of the bacterial cell wall compound peptidoglycan in Gram-positive bacteria (Book 1, Section 3.3). The precise mode of action is via inhibition of the enzyme that catalyses this cross-linking process – a transpeptidase. Recall that a large component of the cell wall of Gram-positive bacteria is constructed from peptidoglycan. Gram-negative bacteria have far less peptidoglycan in their cell walls and indeed penicillin type antibiotics have little effect on them.

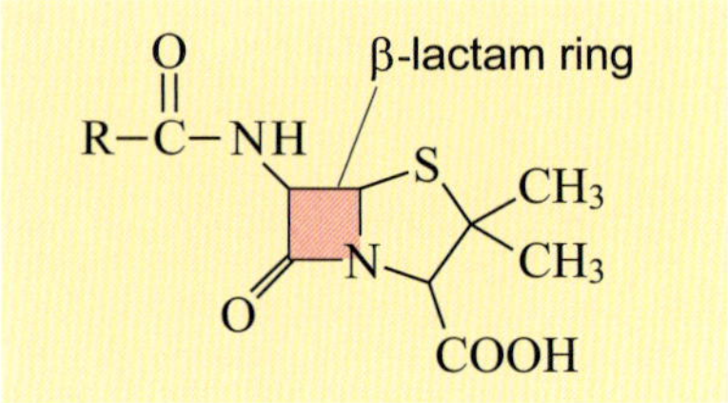

In Gram-positive bacteria, the effect of penicillin is to stop bacterial cells dividing, as they are unable to synthesise new cell wall material. It is, therefore, most effective on sensitive bacteria that are actively dividing and therefore have a high requirement to synthesise peptidoglycan; it has much less effect on stationary phase cells. Penicillin was the first β-lactam type antibiotic; this means it contains a **β-lactam ring** in its molecular structure (Figure 3.8) which inhibits the cross-linkage of peptidoglycan polymer chains in the bacterial cell wall.

Penicillin, a product of fungal species of the genus *Penicillium*, was identified in 1928 by Alexander Fleming but not used with patients until the late 1930s and early 1940s. Yet as early as the mid-1940s, only a few short years later, it was clear that resistant strains of *S. aureus* were emerging.

There are now other types of antibiotics available, including many synthetic derivatives of penicillin, but despite significant investment in this area by the pharmaceutical industry, it is proving impossible to produce antibiotics that remain effective over a sustained period against many pathogenic bacteria. At present antibiotic-resistant strains of bacterial species including *S. aureus* are the cause of a significant number of life-threatening infections all over the world. Staphylococcal infection is a serious problem in hospitals and other clinical environments; anyone in hospital for a surgical procedure is concerned about the risk of acquiring a difficult to treat infection. Even more worryingly, the incidence of so-called community-associated strains is increasing globally. Throughout all countries there has been a rise in the number of patients recorded with a staphylococcal infection, usually a soft tissue or skin infection that has not resulted from a hospital stay. This worrying development suggests that resistant and extremely pathogenic strains of *S. aureus* are becoming commonplace in the general community.

One such antibiotic-resistant strain is **MRSA** – methicillin-resistant *S. aureus* (Figure 3.6). Methicillin is one of the large range of penicillin-type antibiotics. It was developed in the 1960s in response to strains of *S. aureus* found to be resistant to 'ordinary' penicillin. Resistant *S. aureus* strains can synthesise **β-lactamase** (sometimes called penicillinase), an enzyme that can break down penicillin, conferring resistance on the strain carrying the gene for this enzyme. Methicillin is not degraded by β-lactamase and neither are any of the newer types of β-lactam antibiotics now available (e.g. flucloxacillin, dicloxacillin). Although methicillin and its derivatives were useful for a time, further new strains of *S. aureus* soon emerged that were resistant to

Figure 3.8 The chemical structure of penicillin with the β-lactam ring highlighted.

methicillin and all other β-lactam type antibiotics. This new type of resistance results from a change in the structure of the penicillin binding proteins (PBP) found on the surface of the bacterium. This prevents the antibiotic binding to the PBPs and so exerting any effect on the cell. As a consequence, MRSA continues to build and cross-link its cell wall even in the presence of these antibiotics. Although methicillin is no longer used clinically, it is useful in the laboratory to identify strains of *S. aureus* that are resistant to all penicillins.

The bacterial gene for methicillin resistance is called *mecA* and it is located on a mobile genetic element called SCCmec (**s**taphylococcal **c**assette **c**hromosome **mec**). SCCmec acts as a transposon (Book 1, Section 5.8.2), with the ability to insert itself into the bacterial DNA and excise itself later on, to be passed on to other *S. aureus* cells in the population. It is also possible that bacteria from different but related species can transfer SCCmec horizontally between them (Book 1, Section 5.7.5); there is evidence of methicillin resistance transfer from the relatively benign *Staphylococcus epidermis* to *S. aureus*. The potential transfer of virulence factors between species may have huge implications for the worldwide spread of multiresistant pathogens.

Any methicillin-resistant strain of *S. aureus* is now termed MRSA. This doesn't mean that there are no other antibiotics effective against this strain; in fact, there are several that are useful (e.g. vancomycin). One important issue is that, to treat MRSA, these antibiotics cannot be taken orally; they can only be administered either intravenously or, in the case of vancomycin, by painful intramuscular injection. At the time of writing (summer 2012) newer treatments were in the pipeline; some new drugs had been recently licensed for use in the USA and UK. One of these is daptomycin, a lipoprotein antibiotic that interferes with bacterial cell membrane function in Gram-positive bacteria. You will return to the subject of the synthesis of novel antimicrobial compounds in Chapter 4, where you will consider in more detail the properties of these extremely useful chemicals.

Biofilm synthesis

Some bacterial species are capable of forming **biofilms**, aggregations of sessile (stationary) cells embedded in a polysaccharide extracellular matrix (synthesised by the cells themselves) and firmly attached to a surface (Figure 3.9). The surface can be artificial or composed of living tissue but it is usually in contact with an aqueous environment. Indeed many bacteria can 'hang on' to a surface when even very fast-flowing liquids are passing across them. Dental plaque is a well-known example of a biofilm in animals, but biofilms exist in many environments where they provide protection and a significant survival advantage for the embedded cells.

The ability of some strains of *S. aureus* to establish biofilms on various living tissues and on implanted medical devices such as catheters and artificial joints contributes to their pathogenicity. Biofilm-producing strains of *S. aureus* possess specific genes for biofilm-related virulence factors – molecules that allow the bacteria to interact with the surface, maintain the biofilm and also release cells from the mature biofilm in order that new areas can be colonised.

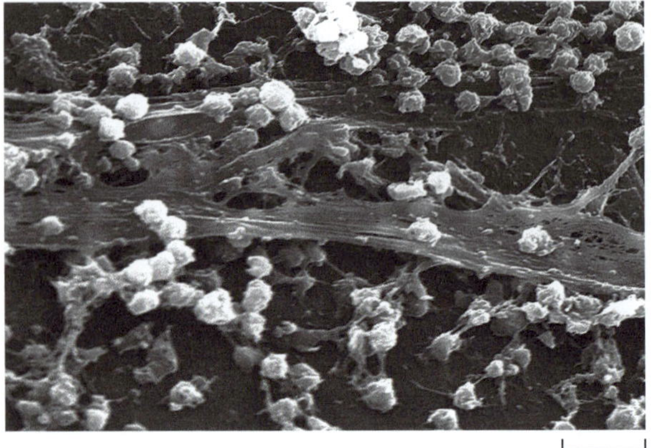

Figure 3.9 An electron micrograph of *S. aureus* bacteria found on the inner surface of a catheter. The sticky-looking substance in which the round cocci bacteria are embedded is a polysaccharide matrix; this gives them protection from antimicrobial agents.

The first stage of biofilm formation (Figure 3.10) involves attachment to the solid surface; this process is once again made possible by adhesins; the bacterial surface molecules commonly found on pili (Section 3.2.2). Secondly, the biofilm matures forming a structure composed of many layers of cells and secreted extracellular matrix. Individual cells within the developing biofilm are capable of communication. As the biofilm begins to form, the *S. aureus* cells synthesise diffusible signal molecules called autoinducers. The concentration of the autoinducer molecule increases around the cells slowly as the number of cells in the biofilm increases. Once a critical concentration of the autoinducer is reached, then receptors inside all the cells within the developing biofilm are activated and express different genes that then control and promote the maturation of the biofilm.

The type of intercellular communication required to promote biofilm formation and maturation is an example of quorum sensing; whereby activity is only triggered when cell numbers have reached a certain threshold density. This permits the behaviour of a population of bacteria to be coordinated for the

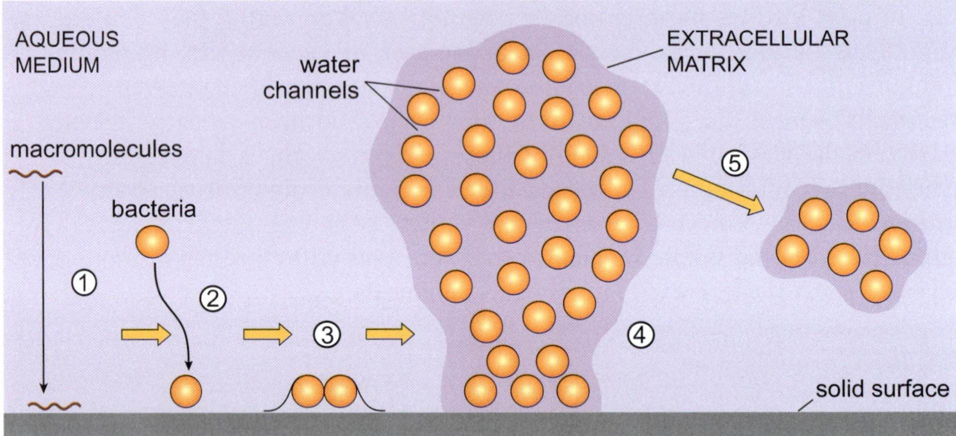

Figure 3.10 The formation of a biofilm on a solid surface such as a catheter: (1) surface conditioning – the surface becomes coated with macromolecules from the aqueous medium; this favours bacterial adhesion; (2) docking – primary bacterial adhesion; (3) locking – secondary bacterial adhesion; (4) biofilm growth and maturation – the bacterial cells are surrounded by a extra cellular matrix; (5) biofilm dispersal.

benefit of the whole population rather than the individual bacterium; you met another example of this in Chapter 2, Section 2.3.5 with bioluminescent bacteria. Intercellular communication to promote biofilms allows the bacteria to successfully colonise and persist in a wide variety of environments. *S. aureus* in biofilms typically shows greater resistance to antimicrobial chemicals, including antibiotics, than free-living cells; possibly because the cells lower down in the biofilm are protected locally from these agents. Similarly where biofilms form inside host tissues, the presence of the biofilm protects the bacteria from the host's immune system. Tackling biofilm formation on equipment and instruments used in invasive surgical procedures is hugely problematic in hospital settings. For example, biofilms are detected in approximately 80% of all infected replacement joints. Pacemakers, catheters and heart valves can all become contaminated with biofilms and in many cases the only treatment is to remove the infected device and replace it.

On a more positive note, the knowledge that these important pathogens are quorum sensing provides a possible way of controlling these infections. You will investigate an example of harnessing the quorum sensing system as a basis for novel therapeutic drugs in Chapter 4. Now that you are familiar with the concept of virulence factors and antibiotic resistance you will have the opportunity, in this first case study (Activity 3.2), to further investigate the impact of pathogens such as MRSA on human populations.

Activity 3.2 Case study 1: MRSA – conquering the superbug

(LO 3.1) Allow 60 minutes to access the online resources, read the associated notes and answer the questions in this activity

In this activity you will access a number of up-to-date online resources to allow you to investigate the pathogenicity of MRSA.

Summary of Section 3.2

- Bacterial pathogens bring about disease by invading the tissues of their host.

- Vertebrates possess immune defences that protect against pathogens; these defences can be innate or adaptive.

- Effective, innate defence mechanisms against bacteria include the use of lysozyme, the alternative complement system and cell-mediated processes.

- Different strains of bacterial species are characterised by their virulence: virulence factors include production of exotoxins and the presence of adhesins.

- Enteropathogenic *E. coli* is an example of a strain of bacteria capable of causing acute disease due to expression of specific virulence factors associated with attachment to the host's gut lining.

- MRSA is a highly pathogenic strain of *S. aureus* that possesses a range of virulence factors including resistance to antibiotics (β-lactam type antibiotics and methicillin) and biofilm formation.

- MRSA demonstrates resistance to β-lactam type antibiotics due to modified penicillin binding proteins on its surface.

- The formation of biofilms by certain species of bacteria presents an infection threat in health care situations; quorum sensing allows bacteria within a biofilm to communicate and express en masse genes that will help mature and sustain the biofilm.

3.3 Subversion of cellular processes by viruses and intracellular parasites

Bacteria generally damage cells from the outside, whereas viruses bring about damage from inside the cell. As obligate intracellular parasites, viruses have a variety of strategies for subverting normal cellular processes and directing them to their own ends. This disruption of cellular activities can often lead to cell death and the consequent destruction of tissues in infected organisms. In this section you will study the effect of viral infection on cells in general and then examine a particularly important viral pathogen in animals: foot-and-mouth disease virus. All viruses are obligate intracellular parasites; they only show any sign of 'life' when they are infecting a living cell. In fact, it has long been debated whether or not viruses should be considered living organisms: many scientists consider them to have too few characteristics of 'life' to be strictly alive, whereas others consider them to be very highly evolved parasites. There are also non-viral infectious agents that can enter cells and disrupt normal cellular processes. *Mycobacterium tuberculosis*, the pathogen responsible for tuberculosis, is a classic example of a bacterial species that can invade cells; in this case in the lungs, bringing about massive damage and ultimately cell death. *Mycobacterium* is classed as a facultative intracellular parasite as it is capable of 'living' independently, but this mode of infection is relatively rare amongst bacteria; it is the viruses as a group that are the dominant intracellular parasites. You will begin by considering the general characteristics of viruses.

3.3.1 Viruses: general characteristics

Mature virus particles, also called virions, consist only of a small amount of genetic material, which is complexed with proteins and packaged into a protein coat called a capsid.

Viruses infect the cells of animals, plants, protists, fungi and bacteria. Viruses that infect bacteria are known as bacteriophage (meaning 'bacteria eaters') or phage, for short. So far, more than 3600 types of virus have been described in detail, which is probably only a small fraction of the total number that exist, and new viruses continue to be described. Viruses can be seen only with an electron microscope, and the first electron micrograph of a virus was made in 1941, about 10 years after the first electron microscope was built. Figure 3.11 shows some examples of the diversity of virus morphology.

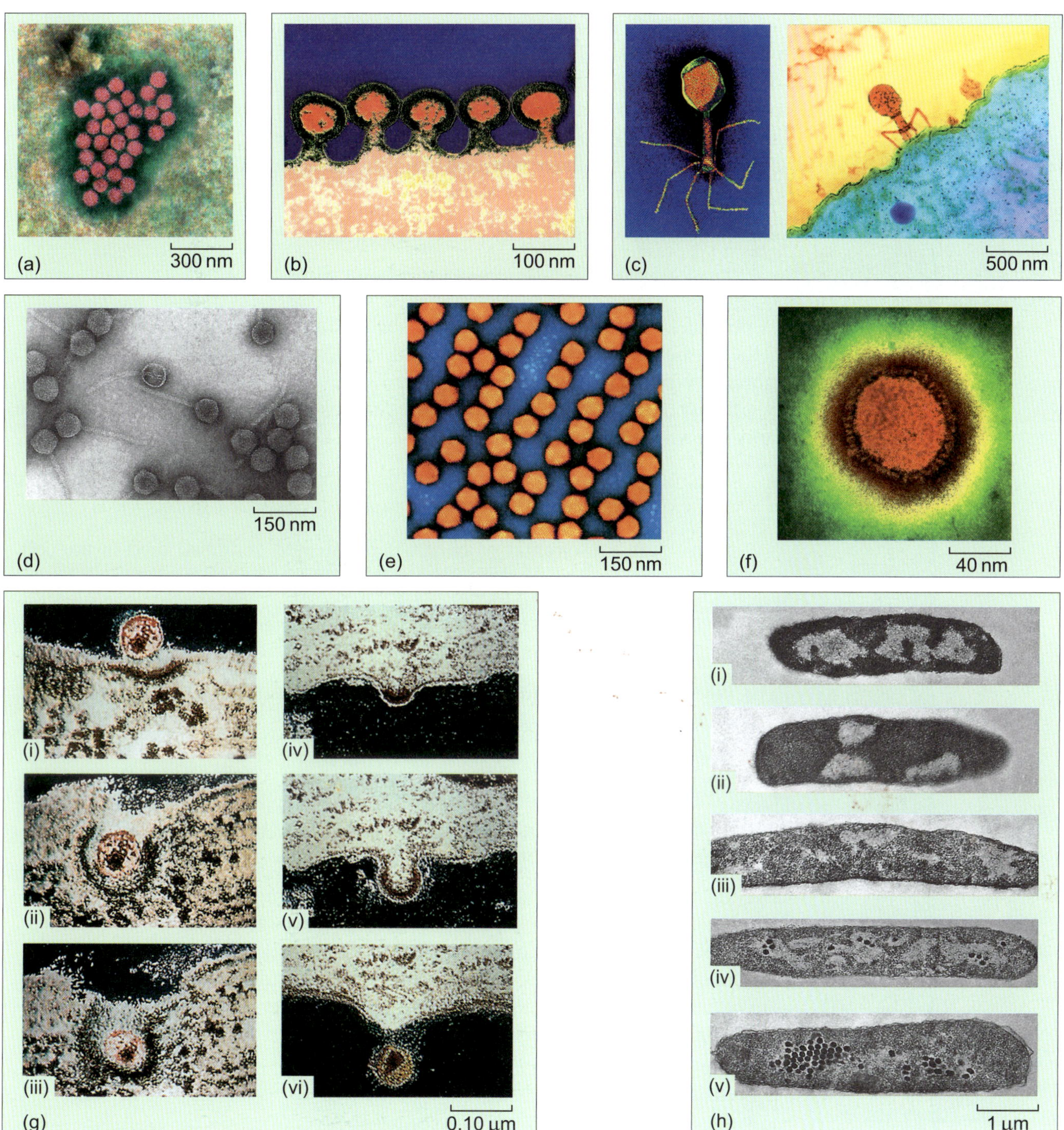

Figure 3.11 Diversity of viruses. (a) Human wart virus (many virus particles). (b) HIV virus budding from the surface of a cell. (c) T phage of *E. coli*; the complex tail structure (left) mediates attachment to the bacterium (right). (d) Phage λ (lambda) of *E. coli*. (e) Adenovirus. (f) Influenza virus. (g) Sequence of micrographs showing: (i), (ii), (iii) the entry of human immunodeficiency virus into a leukocyte; and (iv), (v), (vi) the exit of a newly synthesised virion. (h) The course of a phage infection in *E. coli*: (i) uninfected cell; (ii) 4 min after infection – note the changes in DNA distribution; (iii) 10 min after infection; (iv) 12 min after infection – the first new phage particles can be seen inside the host cell; (v) 30 min after infection, just before lysis – many new phage particles are visible.

Viruses have no cytoplasm, nor any endogenous metabolic activities. They do not grow or undergo division, and the term 'growth' should not be used to describe viral replication. No known virus has the genetic or biochemical potential to release the energy necessary for driving biological processes such as the synthesis of macromolecules. Viruses are therefore absolutely dependent on the host cell for these functions and they often, but not always, kill the cells that support their replication.

■ Suggest how viruses can be studied in the laboratory.

☐ They must be replicated inside cultured cells.

In practice, viruses can therefore only be studied effectively if suitable host cells can be grown in culture. Bacterial viruses are propagated in cultures of laboratory bacteria; plant viruses are grown in either the cells of whole plants, or plant cells in culture. Animal viruses are usually grown in layers of animal cells known as tissue culture.

■ Assuming a virus killed the cells that support its replication, what would you expect to see in a layer of animal cells exposed to this virus?

☐ You would see circular patches where infected cells had died in the cell layer.

The areas of lysis (cell death) caused by a viral infection of a layer of cells are called **plaques** (Figure 3.12).

The function of the capsid of a virus particle is to protect the viral genome from:

• physical damage, such as shearing by mechanical forces

• agents of chemical damage, such as UV irradiation from sunlight

• enzymatic damage, such as by the action of nucleases derived from dead or leaky cells or secreted by animals as defence against infection.

Despite this some viruses are very fragile, for example, HIV is unable to survive outside the protected host cell environment for long. Others, like the smallpox virus, are able to persist for long periods, in some cases for years, depending on the structure of the capsid. The capsid (and, where present the envelope) also play a role in recognition of the host cell and initiation of infection.

Typically a viral capsid is constructed from many identical subunits of one or two kinds, which give the virion a highly symmetrical shape. Virus capsids fall into three main structural types, based on the kind of symmetry they show; this is essentially a result of the way in which the subunits are arranged, either in a rod-like helical arrangement, like the tobacco mosaic virus (TMV, Figure 3.13), or in an icosahedral arrangement, like the adenovirus or the influenza virus (Figure 3.11e and f). The third class includes members of the pox virus family and the T (tailed) bacteriophages of *E. coli* (Figure 3.11c) and has a more complex arrangement. T phages have a combination of structures with an icosahedral head, a helically arranged tail and sometimes tail fibres at its base.

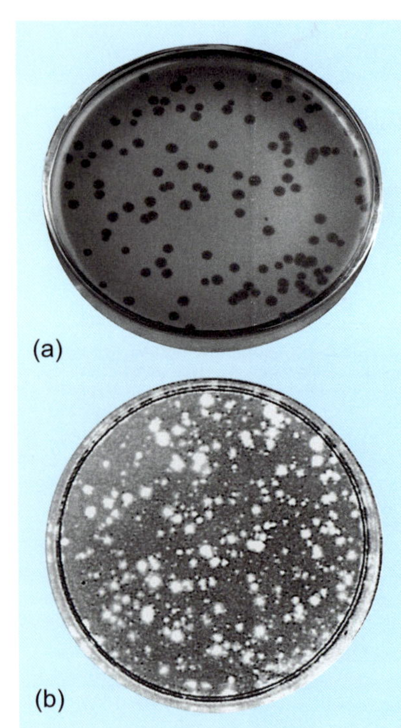

(a)

(b)

Figure 3.12 Plaques of lysis due to viral infection of (a) bacteria growing on the surface of an agar plate and (b) a monolayer of animal cells. Each plaque represents an area of lysed cells arising from one originally infected cell.

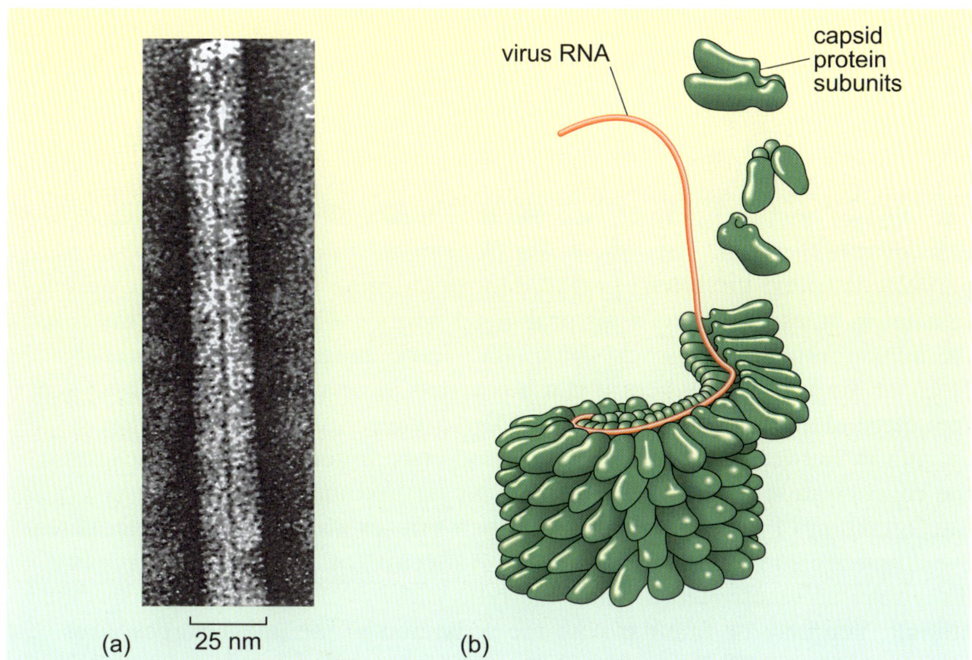

Figure 3.13 Helical symmetry shown by tobacco mosaic virus (TMV).
(a) Electron micrograph of part of a TMV particle. (b) Assembly of a TMV
virion. The RNA genome is arranged helically around a hollow core, and is held
in this configuration by the assembly of identical protein subunits.

Some viruses, such as influenza (Figure 3.11f) are **enveloped** – they possess
an additional membranous structure outside the capsid. This membrane is
derived from the infected cell; it is actually host cell membrane that is
removed with the virus as it leaves the infected cell by a process called
'budding'. Figure 3.11g (iv–vii) shows this process for the exit of newly
synthesised HIV virions from an infected white blood cell or leukocyte.

The majority of viruses can be fitted into one of the three structural classes
outlined above: those with helical symmetry, icosahedral symmetry, or
enveloped viruses based on either of these two.

3.3.2 Virus 'life' cycles

Although the molecular details vary, and in some cases are extremely
complex, viruses have a reasonably straightforward replication cycle. A single
virion attaches to and infects a host cell. Inside the host cell, the viral nucleic
acid becomes exposed and the genome is transcribed and translated by the
host cells gene expression processes to produce viral proteins. Some of these
proteins subvert the host's cellular machinery so that it becomes dedicated
partially or totally to the production of new viral particles. Two sets of viral
molecules are produced: more genome copies and capsid proteins. When these
components are present in sufficiently high concentrations, they self-assemble,
resulting in a large number (in the hundreds) of new virions, which are
released from the cell, usually lysing it. The number of new virus particles
produced is called the **burst size**, and is characteristic of the virus–host
system. The new infective particles can go on to infect further host cells in the

same organism and/or other similar organisms, spreading infection and hence the disease.

There are two timescales for this process. In **lytic viruses**, the whole process happens very quickly and new virions are produced in a matter of minutes or hours, depending on the host cell type. Lytic viruses often cause a complete cessation of host cell growth, because they hijack all the host's resources for their own replication. The time taken from initial infection to release of new particles is called the **latent period**. However, some viruses have the option of remaining quiescent in cells for prolonged periods – many years in the case of the human immunodeficiency virus, HIV. Some integrate their genome into the DNA of the host cell, where it can be carried for many cell generations and replicated along with the host's own DNA; other viruses remain in the cytoplasm but replicate their genomes and coat proteins very slowly so that the concentration of virion components never becomes high enough to trigger the formation of new virus particles. Such viruses are said to be in the latent state, and are sometimes referred to as **lysogenic**, or in a state of **lysogeny**. They may be reactivated to enter the lytic pathway by one or more specific stimuli. You may be familiar with the activation of the latent herpes virus that causes cold sores after you have been out in the sun, exposing you to UV light. These two alternative replication strategies of viruses are illustrated in Figure 3.14.

Viral infections follow three distinct stages:

1 initiation of infection

2 replication and expression of the virus genome

3 assembly, maturation and release of mature virions.

The next few sections will examine each stage in more detail.

The initiation phase

The initiation phase consists of attachment between virus and host cell, penetration of the virus into the cell, and the removal of the viral coat (uncoating). These processes are not necessarily consecutive: for example, T (tailed) bacteriophages (phages) leave all their coat proteins outside the bacterium and inject their DNA into the cell. In this case, penetration and uncoating are simultaneous.

The attachment phase comprises specific binding of a viral attachment protein to a cellular receptor molecule. Virus receptors on cell surfaces may be proteins or the carbohydrate parts of glycoproteins or glycolipids. Some complex viruses (e.g. pox viruses, herpes viruses) use more than one receptor and therefore have alternative routes of uptake into cells. Attachment is an automatic docking process that does not require energy, and receptor binding is controlled by the relative concentrations and availability of the molecules involved. In most cases, the presence or absence of receptors on the surface of host cells determines the type of cell a virus can infect. Therefore, the attachment phase of infection is critical to viral pathogenesis (i.e. whether or not the virus causes a disease).

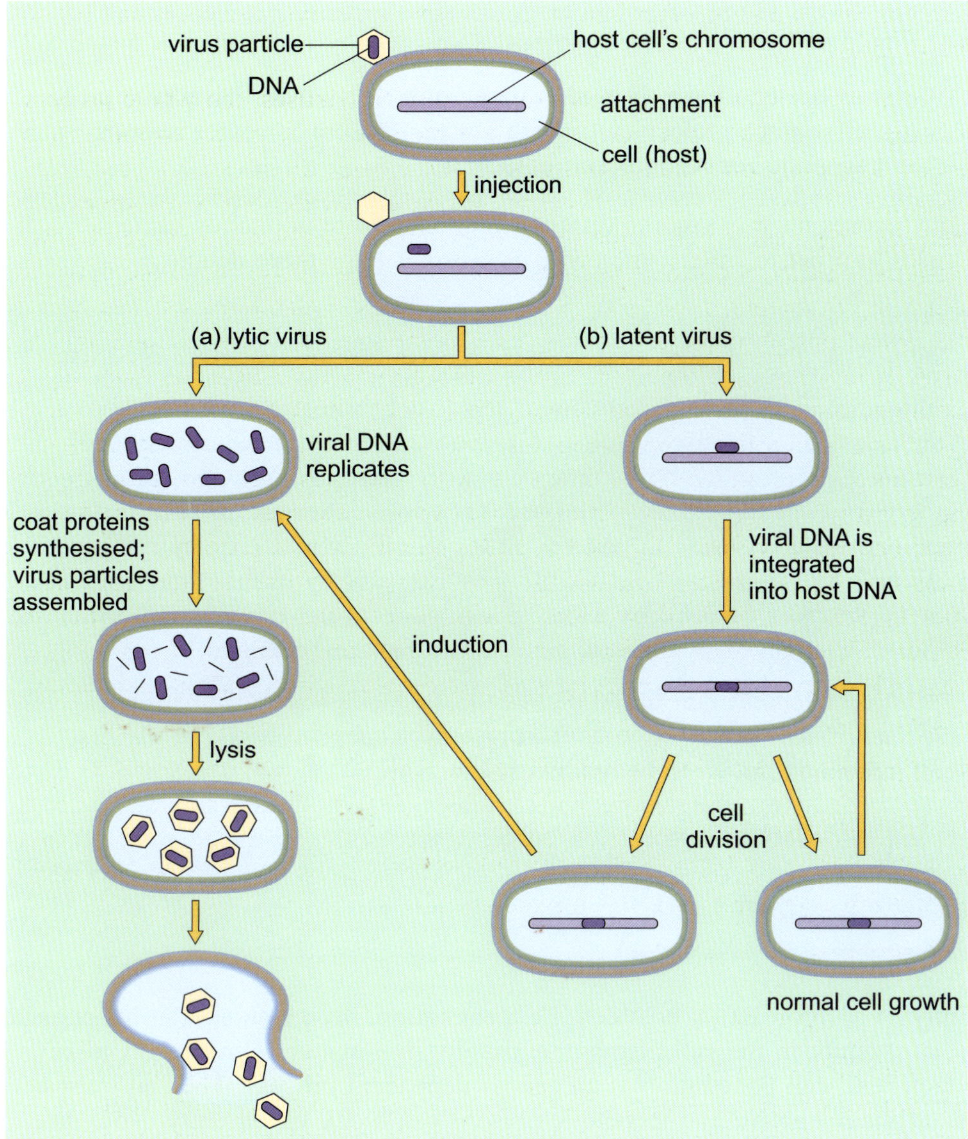

Figure 3.14 Alternative viral replication strategies. Some viruses only ever replicate via the lytic cycle; others deploy both replication strategies, both begin with the entry of the viral DNA into the host cell. (a) The lytic cycle. Incoming viruses are uncoated, and the genetic material is replicated and transcribed and translated straight away. When enough viral components (genomes and coat proteins) are present, self-assembly occurs, yielding a large number of progeny viruses that are released from the host cell and may go on to infect other cells. (b) Latent viruses are not replicated in large numbers straight after infection. They may integrate into the host chromosome where they are replicated along with the host's DNA. A number of environmental conditions may cause them to be induced, when they separate from the host chromosome and enter the lytic cycle.

■ Would cells have evolved molecules specifically for viruses to attach to?

□ Just as bacteria have, it is more likely that the viruses have evolved to exploit surface molecules that are already present for other reasons, such as transporter or recognition molecules.

Plant viruses must overcome different problems from animal or bacterial viruses in order to initiate infection. See Box 3.1 for more details.

Box 3.1 Plant viruses: mechanisms for initiating infection

The outer surfaces of the shoots of plants are protected by a cuticle and the cell wall, composed mainly of cellulose, which surrounds each cell (Book 1, Section 2.4.3). No known plant virus uses a specific cellular receptor of the type that animal and bacterial viruses use to attach to cells; instead, viruses can enter a plant cell only via mechanical breaks in the cell wall. Breaks in plant cell walls are made by herbivorous animals or by enzymes secreted by a fungus associated with transmission of the virus, or simply by accidental mechanical damage to cells.

After replication in an initial plant cell, the lack of receptors poses special problems for the virus in spreading to new cells within the plant. This problem has been overcome by the evolution in plant viruses of specialised viral proteins known as 'movement' proteins which attach to plant cell cytoskeletal proteins, enabling the viruses to travel from cell to cell via plasmodesmata, the cytoplasmic connections between plant cells (Book 1, Section 2.4.3).

Entry of a virion into a host cell, i.e. penetration, occurs soon after attachment to its receptor in the cell membrane, usually in a matter of minutes. Unlike attachment, cell penetration is an energy-dependent process. The cell must be metabolically active and metabolic poisons can block virion entry. Note that whereas attachment is a reversible process, penetration is not. There are three mechanisms by which virions can enter cells: first, endocytosis of the virion into an endosome (Book 2, Section 2.9.2); second, in the case of enveloped viruses, by fusion of the viral envelope with a cellular membrane either directly at the cell surface or, following endocytosis of the virion, with a cytoplasmic vesicle; and third, translocation of the entire virion across the cell membrane.

Once the virion is inside the host cell, uncoating can take place. Uncoating is a general term for the events that occur after penetration, in which the virus capsid is partially or completely degraded and the virus genome exposed, usually still in the form of a nucleic acid–protein complex rather than as nucleic acid alone. The removal of the virus envelope during membrane fusion is the initial stage of the uncoating process for enveloped viruses. In eukaryote hosts, the primary events in uncoating may occur inside endosomes, or directly in the cytoplasm.

■ What environmental conditions in the endosome might contribute to protein breakdown?

☐ Endosomes may fuse with lysosomes, which contain proteolytic enzymes. Lysosomes also have a low internal pH: acidic conditions which would denature many kinds of proteins.

The ultimate product of uncoating depends on the structure of the viral genome. Some eukaryotic viruses such as herpes viruses, adenoviruses and papovaviruses actually retain their capsid after penetration; these are viruses that replicate in the nucleus of their host cell. The capsid proteins of these viruses contain nuclear localisation sequences that attach to the cytoskeleton. This interaction allows the transport of the entire virion to the nucleus: at the nuclear pores, complete uncoating occurs and the viral genome passes into the nucleus.

■ What kind of motor protein might be involved in transporting the virus towards the nucleus of animal cells?

☐ Some kind of dynein may be involved, since these proteins mediate movement inwards from the cell surface (i.e. towards the minus end of microtubules, Book 2, Section 5.2.2).

The replication phase

The replication strategy of a virus depends on the structure and composition of its genetic material. Viral genomes (i.e. the nucleic acid which encodes the genetic information of the virus) are more varied than those of all the prokaryotic or eukaryotic organisms. Unlike the genomes of cells, which are composed of double-stranded DNA, virus genomes may contain their genetic information encoded in either DNA or RNA (TMV, shown in Figure 3.13, is an example of an RNA virus). The nucleic acid comprising a virus genome may be single-stranded or double-stranded and in a linear, circular or segmented form.

Viral nucleic acid is generally short and encodes relatively few proteins (although some viruses with very large genomes containing many genes are known). Some viral genes, the 'early genes', are transcribed as soon as the virus is uncoated, and some of the early proteins thus produced hijack the host's replication machinery to make hundreds of copies of the viral genome. The rest of the viral genome, the 'late genes', encode the late proteins, including the capsid proteins. They are produced in very large numbers too, so that during virion assembly, every copy of the viral genome is encapsulated within a new coat. These processes can be enormously energy-intensive for the host cell, and nearly all its own housekeeping functions may stop as soon as the viral genome is activated.

Remember that viruses are obligate intracellular parasites, able to replicate only inside the appropriate host cells. Therefore the viral genome must contain genetic information in a form that can be recognised and decoded by the particular type of cell infected. Thus the genetic code employed by the virus and the control signals directing the expression of viral genes must match, or at least be recognised by, the host organism.

- What sort of signals would be important for viral transcription?

- Promoters and terminators determine correct transcription. These sequences are known to differ between prokaryotes and eukaryotes (Book 1, Chapter 6) so the virus must carry the appropriate type for its range of hosts.

This specificity means that bacteriophage cannot infect eukaryotic cells, and viruses of eukaryotes are unable to infect prokaryotes. This fundamental distinction between viruses with different host ranges is probably a very ancient characteristic and is likely to have evolved at about the same time as prokaryotes and eukaryotes themselves.

Assembly, maturation and virion release

Assembly involves the coming together of all the components necessary for the formation of the mature virion at a particular site in the cell. The site of assembly varies for different viruses and depends on the site of replication within the cell and on the mechanism by which the virus is eventually released from the cell. During assembly, the basic structure of the virus particle is formed, but this is not necessarily the final structure of the infectious virion.

Some viruses encode proteins that are not structurally part of the virus but whose presence in infected cells is required for normal assembly. Host cell chaperones may also be used during assembly. The symmetry of structural proteins is important in virus assembly, guiding the process and allowing some viruses to self-assemble spontaneously without energy expenditure.

As with entry and uncoating, it is not always possible to separate the assembly, maturation and release of virus particles as discrete processes. It is thought that virus proteins and genome molecules accumulate in the cell until they reach a critical concentration that triggers assembly. In many cases, this concentration is simply based on the amount in the whole cell. Some viruses achieve high levels of newly synthesised structural proteins by concentrating them into semicrystalline arrays, which are visible by light microscopy (e.g. Figure 3.11h).

Maturation is the stage of the replication cycle that prepares the virus particle for the infection of more cells. Maturation often involves structural changes in the newly formed particle, resulting from specific cleavages of viral proteins to form the mature products, or from conformational changes in proteins, such as those brought about by hydrophobic interactions, that occur during assembly.

For lytic viruses, release is a simple process: the infected cell breaks open and releases the virions. Lysis is often caused by the simple disintegration of virus-infected cells, because viral replication disrupts normal housekeeping functions. Many viruses also encode proteins that stimulate apoptosis or programmed cell death (Section 1.3) in their host cells, which can also result in release of virions as the infected cells die.

An enveloped virus can acquire its phospholipid and protein membrane in one of three ways, as summarised in Figure 3.15: (i) as it buds from the host cell directly at the cell membrane, (ii) as it is removed from the cytosol into an intracellular vesicle or endosome, or (iii) as it buds from the nuclear membrane (only possible for viruses that undergo maturation within the nucleus). If many particles are released together, the loss of the membrane may kill the cell or make it leaky.

3.3.3 Viral infections and the immune response

Most viral infections in higher vertebrates, including humans, are usually limited to some extent by the host's immune response, both the innate and adaptive immune systems. A detailed treatment of the complexities of this immune response is beyond the scope of this module however, in order to understand the importance of immunological memory in controlling the spread of some viral diseases, it is useful to outline the role of some aspects of the adaptive immune system.

The immune system can attack viruses before they enter cells but also can recognise infected cells by the fragments of viral proteins which are present on their surfaces. As with defence against bacterial disease (Section 3.2.1) the innate immune system plays a part in the control of viral infections in animals. Virus-infected cells release small proteins called interferons; these are signalling molecules that interfere with viral replication in the host cell. The presence of interferons induces cells to take evasive action, which may include reducing cellular protein synthesis. A type of leukocyte, called natural killer cells, can non-specifically recognise and attack virus-infected cells and the complement system also plays a role in destroying free virus outside the cells. However, there are also several adaptive responses the immune system can mount against viruses; these can lead to the formation of immunological memory, a process that confers long-term protection against a specific pathogen. The most significant adaptive immune mechanisms against viruses are antibodies to combat free virus, and cytotoxic cells to attack virus-infected cells.

The molecules on the surfaces of pathogens (bacteria as well as viruses) are highly antigenic: the immune system is capable of recognising these molecules as 'non-self' and mounting a response against them. The adaptive immune system can recognise each individual antigen specifically as a unique molecular entity, distinguishing one from another with exquisite precision. This property of the immune system is known as antigen specificity.

The mechanism for recognising this vast multitude of different antigens is dependent on a group of leukocytes called lymphocytes. Each lymphocyte has cell-surface receptors that recognise a single type of antigen, or part of an antigen. Within the large, pre-existing population of lymphocytes in every

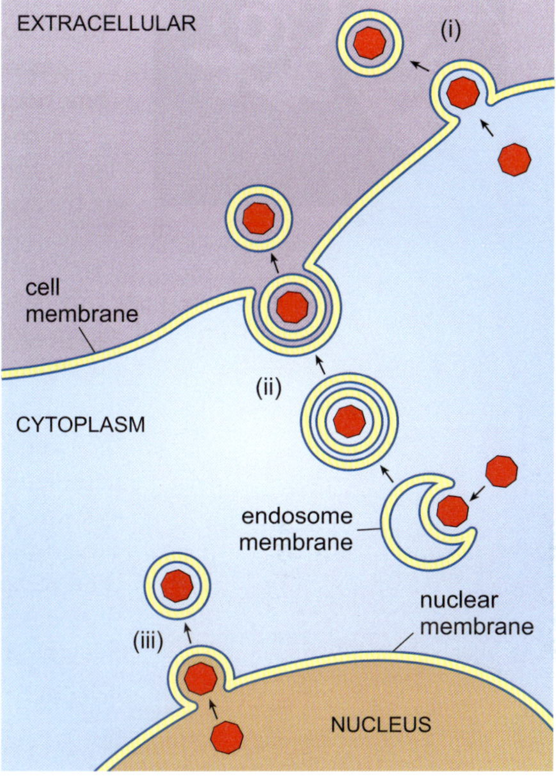

Figure 3.15 Release of budding viruses, showing different sites of budding, where the virus acquires its envelope: (i) cell membrane; (ii) endosome (intracellular vesicle) membrane; (iii) nuclear membrane.

Antibodies are protein molecules able to recognise and bind in a specific manner to target molecules known as antigens (Book 1, Box 2.4).

individual, there are lymphocytes capable of binding to every antigen that exists; this small group of cells is known as a clone.

On infection with a pathogen, members of the clone will encounter 'their' antigen for the first time. They will immediately bind the antigen and undergo 'clonal expansion'. Here the lymphocytes are stimulated to divide frequently to produce a large pool of cells all capable of recognising the same antigen as their original parent lymphocyte. Some of these new cells become long-lived 'memory' cells that remain in the circulation ready to respond rapidly with any subsequent infection by the same pathogen. Others of the expanded clone actively defend (by a variety of methods) the host against the infection and bring it under control.

There are two main types of lymphocyte. **B cells** are activated to divide and become plasma cells that secrete large quantities of antibodies. The antibodies bind to the surface of free virus particles, causing them to clump together so they cannot invade cells and also mark them out for recognition by the complement system and for phagocytosis by leukocytes of the innate immune system. The second class of activated lymphocytes, called **T cells**, recognise virus-infected cells. T cells carry antigen receptors on their surface; they can bind to specific antigens expressed on the surface of virus-infected cells and then destroy the cell. Bacterial infections can stimulate the same types of adaptive response as viruses do.

This part of the immune system is termed 'adaptive' because it adapts by producing an amplified population of antigen-specific immune cells when a particular pathogen (or indeed any antigen) is encountered for the first time. The 'memory' cells that remain after the first infection has been cleared mean that the response to a second or any subsequent invasions by the same pathogen is much faster and more effective than on the first occasion. When a certain kind of pathogen infects its host for the first time, a **primary adaptive response** (often abbreviated simply to primary response) occurs which is relatively slow to develop – it is barely detectable for about 7–10 days and then builds slowly to a peak within 2–3 weeks and lasts a few weeks at most (Figure 3.16a). During this period, it is likely that the host will experience symptoms of infection that gradually get worse before they get better. However, if the same pathogen is introduced a second time, a greatly enhanced **secondary adaptive response** (or secondary response for short) occurs; it develops sooner, lasts longer and displays greater levels of activity than the primary response (Figure 3.16b). The enhanced secondary response to a pathogen may be sufficiently effective to prevent any symptoms from developing, or the symptoms may be much milder than on the first occasion. If subsequent infections do not cause any symptoms, the host animal is said to be immune to that pathogen. This rapid and much more effective secondary response is a result of the expanded clone of lymphocytes which are specific for this antigen.

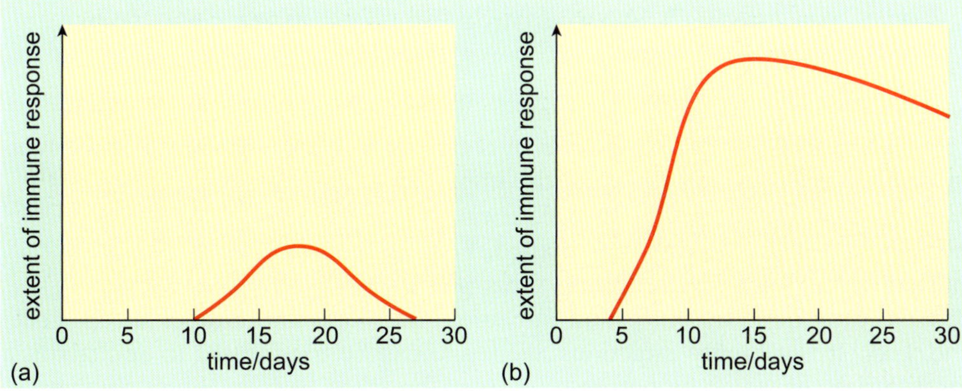

Figure 3.16 The general pattern of (a) a primary adaptive response to a particular pathogen, and (b) the secondary adaptive response to the same pathogen.

Vaccination is one way of artificially inducing immunological memory in individuals to reduce the population of susceptible animals and so limit the spread of a disease. Vaccines contain samples of the unique antigens found on specific pathogens (the vaccine often takes the form of an attenuated strain of the pathogen which cannot cause illness). On injection, the vaccine triggers the primary immune response in non-immune individuals. On any subsequent exposure to the real pathogen, the host will be able to rapidly destroy the pathogen before symptoms develop. The immunity induced in the individual to the specified pathogen can be life-long.

Summary of Sections 3.3.1–3.3.3

- Viruses are obligate intracellular parasites.

- Viral genetic material shows great diversity as it can be DNA or RNA, single- or double-stranded, and linear or circular or in multiple segments.

- Virus structure is based on helical or icosahedral symmetry, but in some cases can be more complex. The virions can be naked or covered with an envelope derived from the host's membrane.

- Viruses replicate by infecting a host cell, and use the cell's molecular machinery to make more viral genomes and capsids. Released virions can go on to infect new host cells, spreading the infection.

- The three main phases of viral infection are: initiation, when viruses gain entry to the host cell; replication, when their nucleic acid genome is replicated and capsid proteins are made and assembled; and release, when the new virions are assembled, become mature and then leave the host cell.

- In vertebrates, including humans, both the innate and the adaptive immune system play a role in limiting viral infections. The adaptive immune system allows the development of 'immunological memory' that can limit the severity of illness experienced on second or further exposure to the same virus. Vaccination is a way of artificially inducing immunological memory.

3.3.4 Case study 2: An example of viral infection – foot-and-mouth disease virus

Now that you are familiar with some of the ways in which viruses subvert cellular processes it is time to look in detail at one specific virus. For the rest of this section, you will focus on the impact of a specific type of virus at the cellular level, the disease symptoms it causes and the possible ways transmission of the virus can be halted. Foot-and-mouth disease (FMD) is caused by foot-and-mouth disease virus (FMDV). This disease is a global, economically important infection in cattle, pigs, domestic buffalo, sheep, goats and yaks. It is a highly infectious agent and can be transmitted to animals in a huge variety of ways: by aerosols (droplets breathed in from direct contact with other infected animals), by contact with contaminated objects such as agricultural equipment or vehicles or from the environment where, unusually for viruses, it can persist between hosts for up to one month.

FMDV causes very distressing symptoms in animals, including severe blistering of the mouth, tongue and hooves. Infected animals have high fever, lameness, excessive secretion of saliva, poor milk production, and lose weight and condition very rapidly. Given the seriousness of FMD symptoms you may be surprised to learn that infected animals do recover from the disease. Only about 1% of infected animals die; however, for those who do not, recovery is prolonged and often results in unresolved health conditions, which reduce the market value of the animal.

Although it is true that during the acute phase of FMD animals are very vulnerable to other infections and indeed in the wild to predation (the vesicles on the feet are so painful, the animal is effectively immobilised), infected animals do regain their health, often within two weeks, as the animal itself naturally limits the infection.

■ Suggest reasons why the diseased animal often recovers from the infection.

☐ The animal's own defence mechanism, the immune system, works to limit the spread of the infection to new cells and to limit virus replication and release.

Unfortunately during the recuperation time the infected animal may pass the infection on to many other animals and the economic repercussions this has, both for the individual farmer and national export of meat and meat products, is very significant. You will investigate the wider economic and social implications of FMD later in this section (Activity 3.3).

The foot-and-mouth disease virus

FMDV belongs to the family Picornaviridae and is classified in the genus *Apthoviruses*. The name Picornaviridae is based on the defining characteristic of this family of viruses, meaning 'small (pico) RNA' virus. Recall that viruses are very diverse in the make-up of their genomes: different virus species can have either single- or double-stranded DNA or single- or double-stranded RNA as their genetic material. In the case of FMDV, the genome consists of a 'sense', i.e. 5′ to 3′, RNA strand (Book 1, Box 6.3 and

Section 6.2.1); also referred to as the (+) sense RNA strand. This means that the infecting RNA as well as being the viral genome can act directly as messenger RNA for translation of viral proteins (Figure 3.17).

The RNA genome is enclosed within an icosahedral capsid with 60 protein subunits (Figure 3.18). Each protein subunit consists of four different viral proteins named VP1, VP2 VP3 and VP4. The virus enters cells by endocytosis (Book 2, Section 2.9.2) and on exposure to the acidic environment inside the endosome, the virus protein coat breaks down, allowing the genome to escape and enter the cytoplasm by an unknown mechanism. Once released, the RNA genome (acting as mRNA) is translated by the host cell ribosomes, which synthesise one large polyprotein product. This initial polyprotein is then cleaved by a virally encoded protease called C3, to produce several viral polypeptides. Some of these are structural, for example, the VP1 to VP4 proteins that make up the capsid subunits. Some are regulatory products or enzymes, for example, the enzyme RNA-dependent RNA polymerase which is required to replicate the viral genome.

■ The infecting FMDV cannot rely on the cell to provide the RNA-dependent RNA polymerase; suggest why not.

☐ Cells have no requirement for this enzyme; in animal cells, RNA is synthesised from a DNA template during transcription and so only requires DNA-dependent RNA polymerase.

The viral encoded RNA-dependent RNA polymerase uses the (+) strand RNA genome as the template for replicating many copies of antisense (−) strand RNA. This (−) strand in turn serves as a template for mass production of new (+) strand genomic RNAs. Each one of these new (+) RNA strands will ultimately be packaged inside a newly synthesised capsid and become a mature virion capable of infecting a new cell. Virus-encoded RNA-dependent polymerase is not a particularly efficient polymerase: it has no proof-reading activity (Book 1, Section 5.3) and consequently makes lots of replication errors. In fact, every new viral genome contains at least one mutation.

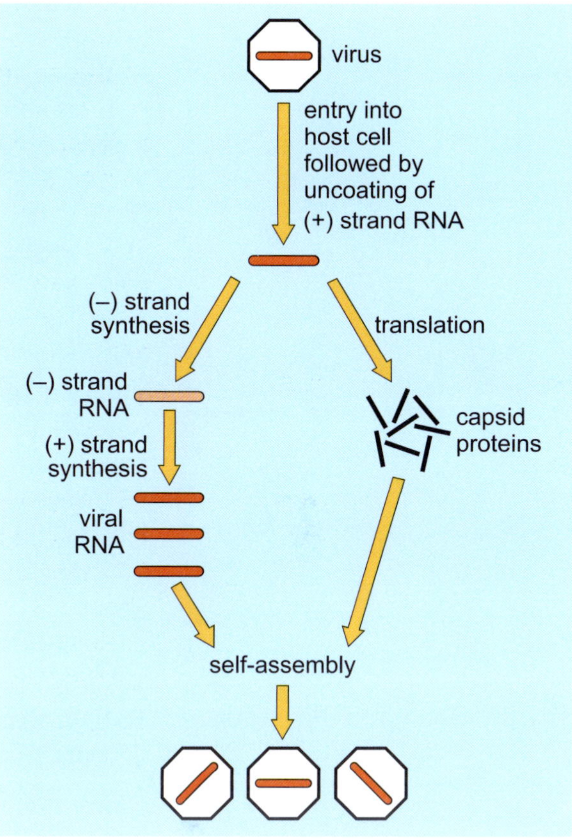

Figure 3.17 Replication of FMDV (typical of single-stranded RNA viruses). Single-stranded (+) virus enters host cell and is uncoated. After uncoating, the (+) strand is used as a template for (−) strand synthesis, and is also directly translated to produce capsid proteins. The (−) strand is used as a template for more (+) strands, and finally (+) strands and capsid proteins self-assemble to form new virions.

■ What implications do these errors in replication have for the genetic make-up of the population of FMDVs?

☐ The population will be genetically very varied; many slightly different strains of FMDV exist even within one infected animal.

This genetic diversity has significant implications for the severity of FMDV outbreaks and the options for treating the infection and limiting its spread.

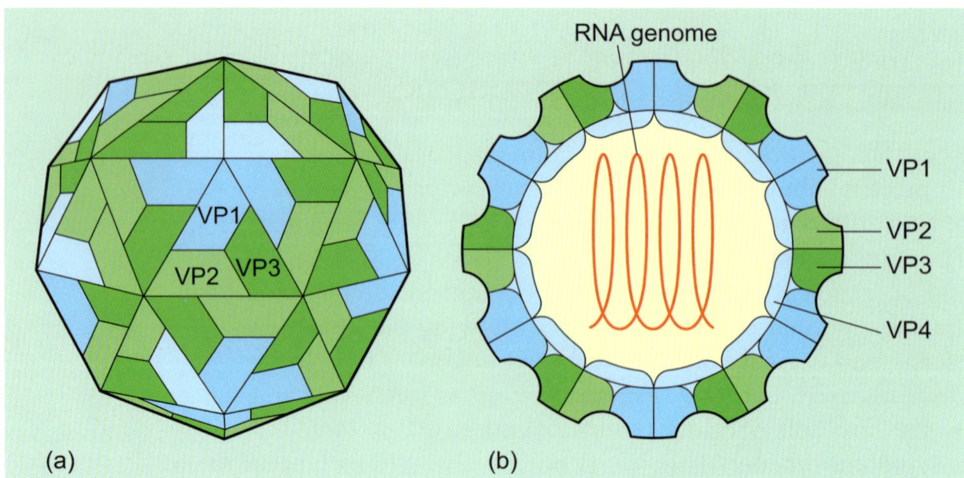

Figure 3.18 (a) External view of FMDV, which is a non-enveloped, spherical virus about 30 nm in diameter, composed of a protein capsid surrounding the naked RNA genome. The capsid consists of a densely packed icosahedral arrangement of 60 subunits, each consisting of four polypeptides, VP1, VP2, VP3 and VP4. VP4 is located on the internal side of the capsid. (b) Cross-sectional view of the virus.

The cellular consequences of FMDV infection

What is happening in the cell while this virus-directed activity is taking place? FMDV, like all picornaviruses, replicates very quickly; it only takes about 8 hours from infection of the cell to the release of hundreds of new virions.

Thirty minutes after entry of the virus, the cell's ribosomes are no longer producing any cellular proteins – this event is called **shut-off** and it is the loss of the cellular protein-making machinery of the cell, which results ultimately in the cytopathic effect (i.e. cell death). To ensure that the host cell ribosomes are only synthesising viral proteins, FMDV deploys sabotage mechanisms to prevent host cell mRNA binding to the ribosomes.

Recall from Book 1, Section 6.6.2 how, under normal conditions, translation of host cell (eukaryote) mRNA relies on certain initiation factors that work in tandem with the ribosome, allowing it to bind to a specific point at the 5′ end of the mRNA strand, the 5′ cap. FMDV disrupts ribosome binding at the 5′ cap of host cell cellular mRNA. FMDV synthesises (via the translation of the viral genome on the host cell's ribosomes) a viral-encoded protease which breaks down the initiation factor associated with the 5′ cap on host cell mRNA (known as the cap-binding complex) and as a consequence prevents any host mRNA binding to the ribosomes.

You may wonder why the FMDV mRNA is not affected in the same way. Actually FMDV mRNA doesn't need the 5′ cap because it relies on different method of binding to ribosomes. It uses a so-called internal ribosome entry site or IRES situated on the 5′ end of its RNA. This different mechanism is not affected by the viral protease and consequently translation from viral mRNA can proceed rapidly. Thus the virus is able to commandeer the cell's own protein-making machinery and completely divert it to its own purposes.

■   What is the method influenza virus uses to ensure that host ribosomes bind to its RNA rather than host cell mRNA?

☐   This virus steals preformed caps from existing host cell mRNA molecules – so-called 'cap-snatching' (Book 1, Section 6.6.1).

In this way the ribosomes become fully occupied translating only viral RNA. From now on, the process is identical to normal host cell protein synthesis. The ribosome travels along the mRNA 'scanning' until it reaches the first AUG codon, after which, translation of the mRNA commences and viral proteins are synthesised rapidly.

Around 3 hours after the start of infection the cytoplasm of the FMDV infected cell begins to appear vacuolated – large membrane bound spaces appear within the cytoplasm. By 4–6 hours the newly assembled viruses are visible in the cytoplasm. Between 6 and 8 hours the cells lyse, releasing the virus. Lysis probably results from shut-off of host cell protein synthesis rather than an event directed by the virus. It is important to realise that not all host cells are susceptible to the virus; as with all viral infections, only certain cells in the animal are targets.

The lysis of groups of infected cells in these tissues results in the painful blisters or vesicles characteristic of the disease (Figure 3.19). As the vesicles burst, huge quantities of virus are released ready to infect new hosts. Transmission of the virus from one host to another may be by a variety of ways, but aerosol dispersal of virus from the mouths of infected animals is probably a highly significant method.

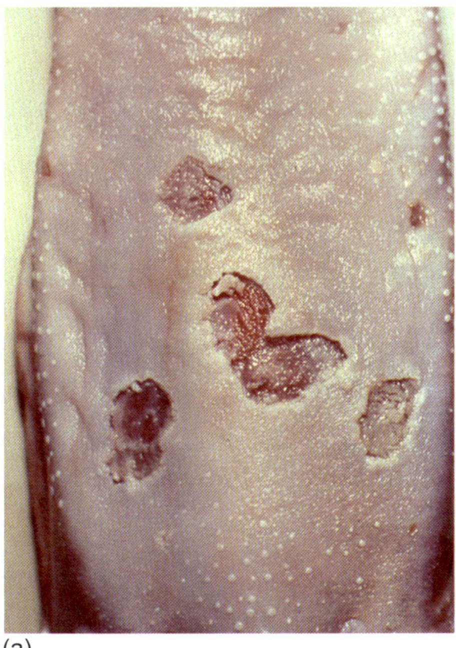
(a)

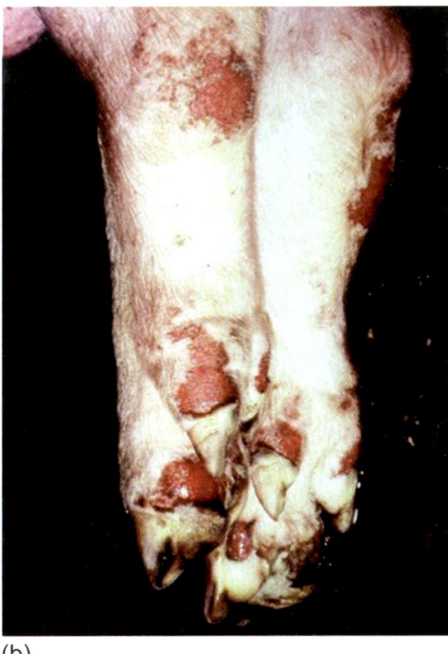
(b)

Figure 3.19 Vesicles on (a) the tongue of a cow and (b) the feet of a pig infected with FMDV.

■   Suggest why cells in the tissues of the mouth and feet, but not other tissues, are vulnerable to infection by FMDV.

☐   Cells in the tissues of the mouth and feet may express certain receptor proteins that are recognised by the virus.

Tackling foot-and-mouth disease

The economic importance of foot-and-mouth disease means that new ways to combat the disease are constantly being sought. In most countries of the world, there are two commonly used methods of handling outbreaks of FMDV infection. One method, frequently adopted in some European countries, is mass slaughter of infected animals followed by stringent quarantine measures. The distressing sight of massive funeral pyres of burning livestock in Europe has been relatively common in the last few decades but in other countries, mass vaccination of susceptible animals is the control method of choice. Why the difference, and are there any other alternative approaches to the containment of this disease?

Unfortunately, vaccines are not an ideal solution for all infectious diseases. To deploy an effective vaccination programme against any pathogen, there needs to be certainty that the vaccine recognises antigens that are consistently present. If the pathogen population has a variety of surface antigens (these are known as different serotypes), then one vaccine that recognises only a single antigen will not be enough. FMDV has at least seven different serotypes and within each serotype there is a spectrum of strains that can be grouped together according to their genomic relationship.

RNA viral genomes are highly variable. Recall that viral RNA polymerases lack a proof-reading capability, so a lot of copy errors are introduced during viral replication. Such replication errors result in progeny viruses with a huge diversity of genome sequences. The sequence of the *VP1* gene (a gene encoding one of the four viral capsid proteins) differs by as much as 50% between different serotypes of FMDV. This results in the expression of significantly different viral proteins, leading to variation in the surface antigenicity of different viral serotypes and therefore a requirement to produce a specific vaccine against each specific serotype.

In addition to the development of effective vaccines, attention has also turned to the use of antiviral drugs. Significant work is underway to identify drugs capable of targeting processes unique to Picornavirus replication. Promising current lines of investigation include inhibiting the virally encoded C3 protease and using RNA interference (Book 1, Box 6.3) to inhibit viral gene expression.

Now you have developed an understanding of the basics of FMDV biology, you will further investigate in Activity 3.3 some recent *epizootic* trends (analogous to the term 'epidemic' when used to describe a human disease) of foot-and-mouth disease, the methods by which scientists have attempted to control the disease and the lessons that have been learnt for the future control of outbreaks.

Activity 3.3 Case study 2: investigating the impact of FMDV

(LOs 3.1, 3.2, 3.3 and 3.5) Allow 90 minutes to access the online resources, video and animation read the associated notes and answer the questions in this activity

In this activity you will access a number of current online resources to allow you investigate the impact of recent FMDV epizootics and the future control of the virus.

Summary of Section 3.3.4

- Foot-and-mouth disease virus (FMDV) causes foot-and-mouth disease (FMD) in livestock.

- FMDV is a non-enveloped, icosahedral (+) sense single-stranded RNA virus.

- There is great genetic variability in the population of FMDV virus.

- FMD is a very significant global problem with epizootics causing wide-scale economic difficulties for affected countries.

- Culling is the current control measure of choice for FMD-affected livestock in many parts of the world. Vaccination may offer an alternative solution, although there are issues related to the existence of multiple viral serotypes.

- Development of effective antiviral drugs may help combat the disease in the future.

3.4 Inherited disease – inherited not 'caught'

So far in this chapter you have concentrated on infectious disease, where a close association between a host and an infecting pathogen causes cell death and damage, either at an extracellular level for most bacterial pathogens or at an intracellular level for viral pathogens. In this section you will investigate human inherited diseases. These diseases really do come from 'within' and demonstrate what can happen when vital genetic information becomes altered.

In Book 1, Chapter 6, you were introduced to the fundamental relationship between DNA and protein: there is a linear relationship between the sequence of base pairs within a gene and the sequence of amino acids within the protein that the gene encodes. A mutation in a gene sequence may result in the inability to synthesise the protein product, or the synthesis of a protein that does not have normal function. This inability to synthesise a functional protein may lead to deficiency in a cellular process and ultimately to disease symptoms, i.e. 'clinical' phenotype.

In inherited diseases, a defective gene or genes are passed on from parent to offspring. These diseases can be classified into two main groups: single-gene disorders and multifactorial disorders. The inheritance of single-gene disorders is relatively straightforward and patterns of inheritance can be established so predictions can be made about the probability of given individuals inheriting a condition. Single-gene disorders usually result from an inability to synthesise a functional product from a specific gene because the gene is abnormal. Although there are many different types of single-gene disorders, they are all by their very nature relatively rare. Some of the more common include: sickle cell anaemia, affecting 1 in 500 African Americans; cystic fibrosis, affecting 1 in 2500 newborn European Union infants; and Marfan syndrome, affecting 1 in 4000 worldwide.

Much more complicated, and far more common, are those genetic disorders demonstrating multifactorial inheritance. They include many of the debilitating diseases of old age: such as Alzheimer's disease, coronary heart disease and type 2 diabetes. The probability that an individual will develop these complex conditions (i.e. their susceptibility to the disease) depends on the inheritance of combinations of many different gene variants (alleles) from across the genome. These diseases occur with increased frequency within some affected families but without showing any consistent and clear pattern of inheritance. Indeed, whether a given family member inherits a particular condition that 'runs' in their family seems to be a result of not only the gene variants they have inherited, but also of many environmental factors including diet, level of exercise taken, stress and alcohol consumption. Recent research seeks to establish the relative contribution of inheritance and lifestyle in these complex disorders.

3.4.1 Case study 3: Single-gene disorders – xeroderma pigmentosum, Huntington's disease and Marfan syndrome

You will now investigate a few single-gene disorders in a little more detail in order to understand their pattern of inheritance and the effect the defective protein has on cellular processes and ultimately on the phenotype of an affected individual.

Xeroderma pigmentosum

One such disease is xeroderma pigmentosum (XP). This is a very rare inherited disorder (Book 1, Section 5.4.2) affecting approximately 1 in 250 000 people worldwide. The disease is most common in Japan and North Africa. It is characterised by extreme sensitivity to UV light, with any exposure to sunlight leading to abnormal pigmentation of the skin (Figure 3.20) and a very high frequency of skin cancer; although the severity of the symptoms differs dramatically from one sufferer to another.

In XP sufferers the median age of developing the first cutaneous (skin) cancer is 8 years, compared to 58 years in the population as a whole, and the incidence of malignant melanoma, the most serious type of skin cancer, is 1000 times greater than average. Two-thirds of all sufferers die before they reach adulthood and in all cases exposed skin is dry, pigmented and covered in lesions. The only treatment is photo protection (extreme avoidance of all sunlight) and surgical removal of malignant tumours as they occur. In the developed world, genetic counselling is offered to families in which the disease occurs.

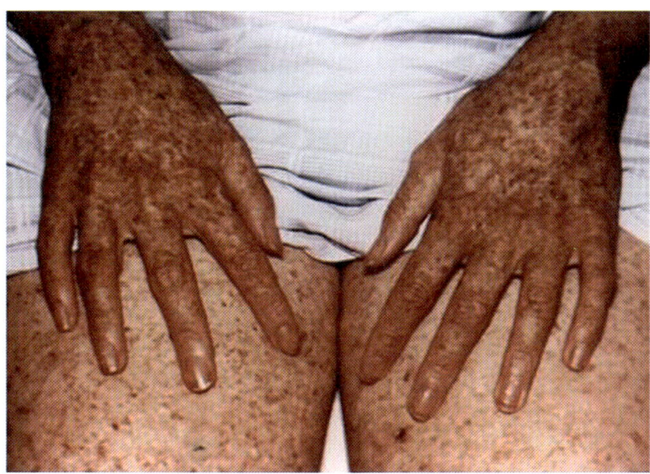

Figure 3.20 Abnormal pigmentation of the skin in a person with xeroderma pigmentosum.

For those with XP the problem is the inheritance of a defective allele coding for one of the proteins involved in the repair of damaged DNA. Recall from Book 1, Section 5.4 that DNA within cells is constantly under attack from a whole variety of agents including electromagnetic radiation, various environmental toxins and even endogenous or 'home-grown' toxic products (e.g. reactive oxygen species, Book 3, Section 2.2.1) from biochemical pathways within the cell itself. DNA must be constantly under repair or the accumulation of these mutations will potentially lead to cancer within affected cells. The action of UV light on DNA causes damage, including the formation of thymine dimers (Book 1, Figure 5.9) that interfere with accurate replication and transcription of the affected DNA.

■   How are thymine dimers and other damaged nucleotides removed from DNA?

☐   These dimers and other damaged nucleotides are removed by the nucleotide excision repair mechanism (NER) (Book 1, Section 5.4.2).

NER is dependent on nine protein products from a number of different genes and individuals with XP can carry defects in any one of these genes. One-quarter of all XP cases have mutations in the so-called *XPC* gene. In unaffected individuals, the role of this gene in DNA repair is to encode a protein that recognises the damaged area of the DNA molecule and binds to it, marking it out for further processing by other gene products of the NER system. In individuals with this type of XP, this protein is defective and so unable to perform this action. The normal (wild type) *XPC* gene contains 16 exons and 15 introns and codes for a protein of 940 amino acids. Pathogenic variants have been found with single base changes leading to nonsense mutations (Book 1, Section 5.6.1).

■   Define the term 'nonsense mutation'.

☐   A nonsense mutation is a mutation that results in a premature stop codon, leading to the synthesis of a truncated protein product.

Hence individuals with a mutated *XPC* gene are unable to effectively repair damaged DNA. The complexity of the NER system, with so many different potential inactivating mutations in the genes encoding the nine different proteins, explains to some extent the variability in the type and severity of XP symptoms from person to person. Another NER mutation causing XP occurs in the *XPG* gene and is responsible for around 6% of cases of XP. The normal product of *XPG* is an endonuclease, one of the enzymes that cleave the DNA chain in order to excise the damaged section. Some individuals with XP who have an *XPG* mutation do express some NER endonuclease, but it is not as active as that synthesised from the normal gene. This is probably a result of a missense mutation (Book 1, Section 5.6.1) in the *XPG* gene.

- ■ Explain why a missense mutation can still produce active product.

- □ A missense mutation is a single base pair substitution leading to the insertion of one incorrect amino acid in the protein product. The effect on the activity of the protein depends on the location of the altered amino acid within it. An altered amino acid may make only minor alterations to the higher-level structure of the protein, allowing it to have some catalytic activity or possibly the activity of the altered protein may even be enhanced.

Individuals with some *XPG* activity do show symptoms of XP but they are not as severe as the symptoms in individuals with nonsense *XPG*, where truncated and completely non-functional endonucleases are produced. In fact, individuals with a nonsense *XPG* mutation frequently have neurological symptoms in addition to skin-related issues, with progressive cognitive impairment, hearing loss and brain abnormalities.

XP therefore demonstrates a huge range of clinical phenotypes dependent on which of many different single genes encoding the NER mechanism is affected and to what extent those mutations reduce expression and/or impair the function of the relevant gene product. Genetic diseases frequently show this enormous range of pathology and only a very few diseases are known to be a result of a specific mutation in only one gene. In contrast, cystic fibrosis is a rare example of an inherited condition where as many as 70% of all those affected possess exactly the same mutation.

You will have the opportunity to learn more about the range of clinical manifestations of XP in Activity 3.4 but before this, it is first important to appreciate how these mutated genes are inherited. XP is inherited in an autosomal recessive manner.

- ■ Explain the term 'autosomal' in this context.

- □ This means that for XP, the mutated genes are carried on the autosomal chromosomes and not on the two sex chromosomes, X and Y.

- ■ Explain the term 'recessive' in this context.

- □ A recessive allele is one that only shows its effects in individuals who are homozygous for that allele. This means that to have the disease a recessive allele must be inherited from each parent. The parents of an individual with XP, if they are unaffected by the disease themselves, must be heterozygous carriers of the affected gene.

- ■ Why are heterozygous individuals for XP asymptomatic?

- ☐ The presence of one normal gene copy in the genotype allows the cell to synthesise sufficient functional NER proteins.

XP is a disease that is more common in some populations than others, and *consanguinity* may be strongly implicated in families where it occurs. Consanguinity is an indicator for the degree of relatedness of two individuals within a population. A consanguineous marriage implies a union between two people with a high degree of shared ancestry. Consanguinity seems to be a factor in around 30% of reported XP cases. Recessive conditions such as XP can remain 'hidden' in populations for many generations as carriers show no symptoms of disease, but where couples share some common ancestry the chances of a child inheriting a recessive allele from each parent are higher than amongst couples with no shared relatives. Isolated rural populations or cultures where marriage of relatively close relations is favoured, have a high incidence of XP and other recessive inherited conditions.

- ■ Are dominant conditions easier to detect within families?

- ☐ One would expect so, as affected individuals only need to inherit one deleterious allele from one parent to show the effects of the genetic disorder and it is usually clear within the family where the affected individuals are.

You will now consider a dominant inherited condition: Huntington's disease.

Huntington's disease

Huntington's disease (Book 1, Section 4.7.1) is one such dominant disorder, although it isn't as easy to prevent passing on the disease as one would expect. The disease phenotype often doesn't become apparent until middle age when gradual loss of coordination, frequent involuntary movements and cognitive decline occur. The disease is eventually fatal. This late onset means that affected individuals often pass on their dominant allele to their children before they realise they have the disease. Pre-symptomatic genetic testing is now available to detect the dominant allele in an individual's genotype, but this causes huge ethical, psychological and practical problems for affected families. Is it perhaps better not to know that you are going to develop a very distressing and fatal illness in midlife?

Huntington's disease is a **neurodegenerative** condition. Neurons are the highly specialised cells in the brain and nervous system that receive, process and transmit electrical and chemical stimuli. When neurons degenerate in the brain, significant loss of muscle coordination and cognitive function is inevitable. The gene affected in Huntington's disease codes for a protein called huntingtin and is located on the tip of chromosome 4. A **trinucleotide repeat** sequence exists within a particular location in the normal huntingtin gene (*Hd*). This means that a sequence of three nucleotides is repeated over and over again in the gene. In this case, the repeat is the triplet CAG in the non-template strand and GTC in the template strand. In unaffected individuals this stretch of DNA consists of between 11 and 34 repeats of this triplet. In individuals with Huntington's disease this region of DNA is substantially longer, consisting of between 40 and 100 triplet repeats. The more repeats there are, the earlier the onset and the greater the severity of the symptoms of Huntington's disease.

■ Use Figure 5.13 in Book 1 to determine which amino acid is coded for by the mRNA codon CAG.

□ CAG is the mRNA codon for glutamine (Gln).

The presence of the repeated triplet means that the protein encoded by this gene will contain a variable length of repeated glutamines. There will be one Gln in the protein for every repeated CAG triplet in the gene. Repeated glutamines up to around 34 (within the normal range) produce normally functioning protein. However if there are more than 34 glutamines in the huntingtin protein then the individual may suffer the symptoms of Huntington's disease. Sadly, the severity of the symptoms and their earlier onset seems to increase with each generation of an affected family as more and more CAG repeats are accumulated within the *Hd* gene. This is known as **genetic anticipation** and is a common feature of other genetic disorders caused by trinucleotide repeats.

■ Given that Huntington's is a dominant condition and is fatal, one you might expect the disease-causing *Hd* gene be eradicated from the human population. How is it that this allele is not eliminated?

□ Because this is a late onset disease in most cases, the dominant disease-causing gene has been passed on to children, and even grandchildren before any symptoms occur; it is not a condition that affects the reproductive success of the individual.

The mutation resulting in Huntington's disease is a 'gain of function' type (Book 1, Section 5.6.2), whereby the altered huntingtin protein gains toxic effects. The abnormal protein aggregates within neurons, forming visible protein fibres (Figure 3.21a and b). It seems to be toxic to nerve cells in the basal ganglia (Figure 3.21a) and the cerebral cortex and the degree of toxicity is directly related to the number of glutamine repeats. The abnormal protein aggregates within neurons (Figure 3.21b) in these areas of the brain and leads to progressive cell death.

The basal ganglia are the site of voluntary motor control in the brain. The cerebral cortex is the most highly developed part of the brain and is responsible for conscious thought, memory and language.

The reason why these aggregates of abnormal huntingtin protein bring about neuronal cell death is not yet clearly established, neither has the full extent of the role of the huntingtin protein in normal cells. Huntingtin has been shown to interact widely with a variety of other proteins including those involved in cell signalling, transport and transcription. A range of critical defects have been found in the affected neurons of people with Huntington's, including a non-functional proteasome system – the system that tags and disassembles defective proteins (Book 2, Section 1.5.4), reduced metabolic rate and reduced neurotrophin concentrations. Neurotrophins are vital chemicals secreted by neurons, which signal to surrounding neurons preventing them initiating apoptosis. Ongoing research is gradually trying to piece together the

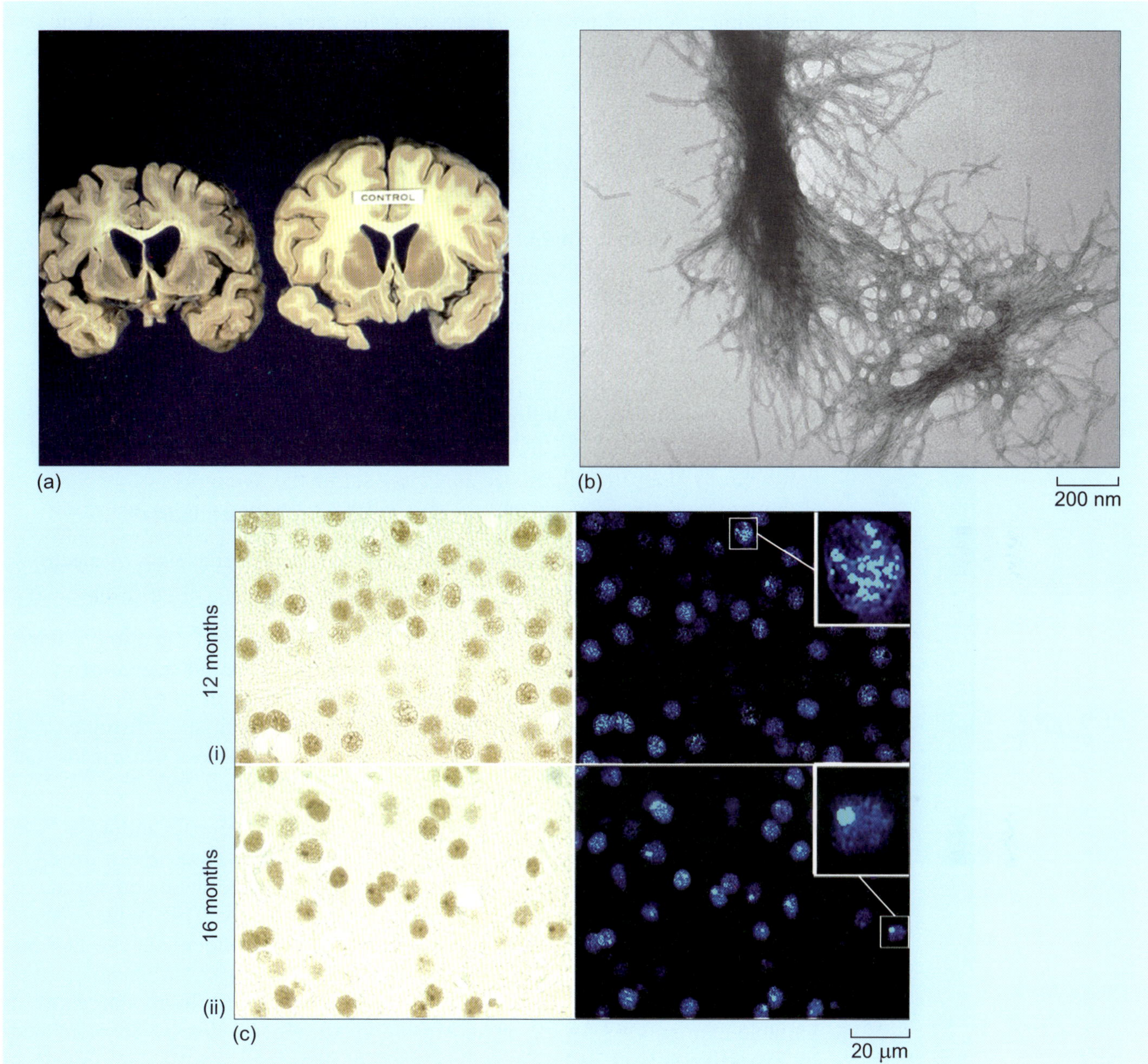

Figure 3.21 (a) The human brain, showing the impact of Huntington's disease on brain structure in the basal ganglia region of a person with the disease (left) and a normal (CONTROL) brain (right). (b) Transmission electron microscopy demonstrates the fibrillar nature of huntingtin aggregates within cells of the human brain. (c) The nuclei of individual neurons in the brains of mice used to model the Huntington's disease processes seen using microscopy in (i) mice that are 12 months old and (ii) mice that are 16 months old. Both left-hand images show neuronal nuclei (here seen as brown spheres) in the cells in the region of the brain that modulates motor activity. These nuclei are expressing a human polyglutamine containing protein, which develops into small protein aggregates. The aggregates are brown in colour due to tagging with an antibody against the human protein. They are more easily seen and quantified using an imaging process (right-hand images) that highlights the aggregates in blue. The aggregates of protein can be seen distributed throughout the nucleus more clearly when enlarged (right, inset). Over a time of 16 months (ii), these small aggregates develop into larger nuclear collections of protein known as inclusions (right inset).

underlying effects of abnormal huntingtin and there is a very long way to go before we fully understand this disease.

There is also some evidence to suggest that huntingtin may have a role in regulating mitochondria. A recent study (Song et al. 2011) has demonstrated that aggregates of huntingtin protein within the nerve cell increase the activity of another protein, DRP1 (mitochondrial fission GTPase **d**ynamin-**r**elated **p**rotein-1). As the name suggests, DRP1 is an enzyme that controls mitochondrial fission. This is the process by which cells manufacture new mitochondria by splitting existing ones, to ensure the cell's complex energy needs are met; newly split mitochondria can migrate around the cell to areas of high energy demand, for example, the synapse of neurons. The increased activity of this protein seems to send mitochondrial fission into overdrive, producing many fragmented, non-mobile and undersized mitochondria in affected cells. Under the influence of overactive DRP1, mitochondria in neurons have been found (using the electron microscope) to have highly disrupted inner membranes.

- What is the function of the inner membrane of mitochondria?

- It is the site of the all-important energy-releasing mechanism of electron transport (Book 2, Section 3.6.3).

Crucially, there is a direct correlation between the degree of mutation within the huntingtin gene (the number of trinucleotide repeats) and degree of mitochondrial defects – this could account for the variation in severity of symptoms shown in this genetic disease. Neural cells with such disrupted mitochondria are very sensitive to apoptosis and it is this loss of neurons that leads to all the typical symptoms of Huntington's disease.

Despite the general correlation between length of CAG repeats and onset and severity of symptoms, two individuals with Huntington's disease can have exactly the same number of CAG repeats in their mutated gene and the age of onset and severity of their symptoms will differ markedly. This suggests that other factors, both genetic and environmental, may be influential. The fact that individuals with the same genotype can demonstrate a different phenotype is called **variable expressivity**. There are other inherited conditions that show variable expressivity, for example, Marfan syndrome – a disease characterised by excessive growth of the skeleton.

Marfan syndrome

Marfan syndrome, like Huntington's disease, is a single gene dominant inherited disorder. The worldwide incidence of Marfan syndrome is estimated to be 2.3 per 10 000 in the population. This condition has a huge range of symptoms including optical, skeletal and cardiovascular abnormalities; although there is a huge range of symptoms, some sufferers seem to be only very mildly affected whilst others are so severely affected that the disease is life-threatening.

The symptoms of Marfan syndrome result from a disorder of the connective tissue (the fibrous tissue that supports all body tissues and organs). The affected gene, *FBN1,* is on chromosome 15 and codes for the glycoprotein fibrillin-1. This protein monomer is a major component of the microfibrils that make up the extracellular matrix of connective tissue. Each fibrillin-1 monomer is synthesised inside a cell and then transported into the extracellular space to bind to other fibrillin-1 monomers and other proteins to produce a microfibril. These microfibrils also associate with elastin fibres within the connective tissue and together these structures provide support and elasticity. In Marfan syndrome, production of fibrillin-1 may be affected in many ways dependent on the specific causal mutation. Production of fibrillin-1 can be reduced, or fibrillin-1 with an impaired structure may be synthesised so it either cannot form microfibrils or it cannot be exported from the synthesising cell. As a result, tissues such as bones, arteries and the tissue of the eye, which normally rely on fibrillin-1 for support and elasticity, are defective. Another function of microfibrils within connective tissue is to hold onto (sequester) various growth factors, rendering them inactive. With decreased microfibril production, these active growth factors are present at increased levels and this alters growth and repair at affected locations. The classic characteristic of individuals with Marfan syndrome is extreme height, with long arms, fingers and toes, the result of the excess growth in the skeleton (Figure 3.22a and b).

There are more than 600 different possible known mutations in the *FBN1* gene that can cause Marfan syndrome. Around 60% of these mutations are simple point mutations where one amino acid in the fibrillin protein is changed. The degree of disease symptoms experienced by the affected individual will depend on the location and type of the mutation in the *FBN1* gene. It is therefore the variability in the type of mutation that leads to the variable expressivity in the phenotype.

Now that you are familiar with the genetic basis of the three single-gene disorders described above, you should complete Activity 3.4. This will allow you to further explore the symptoms, diagnosis and inheritance patterns of these diseases.

Activity 3.4 Case study 3: Human single-gene disorders

(LOs 3.1, 3.6, 3.7 and 3.8) Allow 60 minutes to access the online resources, read the associated notes and answer the questions in this activity

In this activity you will access a number of current online resources to allow you investigate the impact of three single-gene disorders in humans: xeroderma pigmentosum, Huntington's disease and Marfan syndrome. This activity will allow you to determine the pattern of inheritance of these disorders, to practise using human pedigree analysis (Book 1, Section 4.7) and to find out more about how real people are affected by these serious and debilitating diseases.

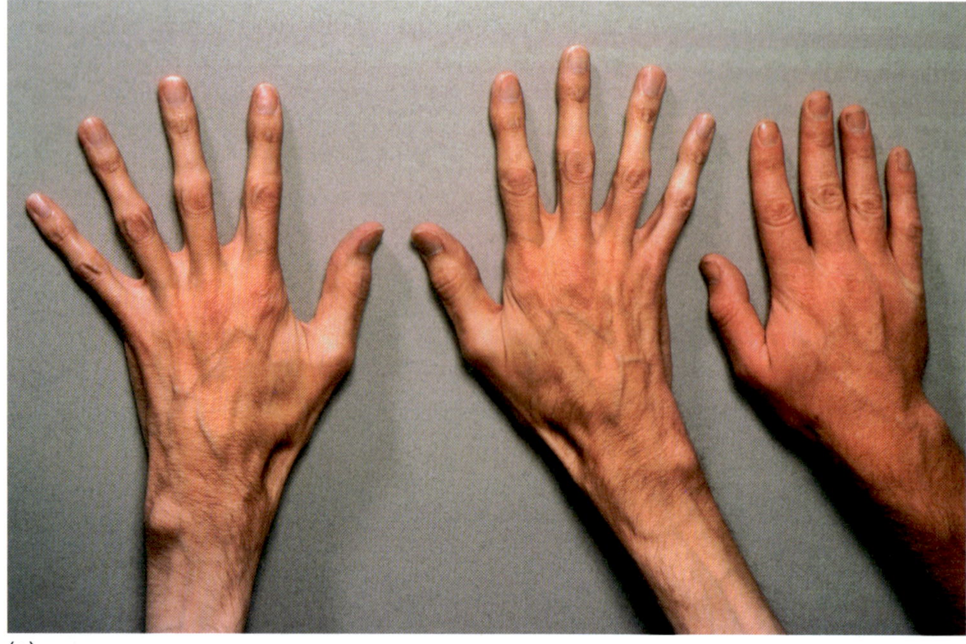

Figure 3.22 (a) Photograph of the hands of a person with Marfan syndrome, showing abnormally elongated and slender fingers. A normal hand has been included in the photo for comparison. (b) Photograph of the arm span of a person with Marfan syndrome – his height is 1.91 m and he has an arm span of 1.98 m.

Summary of Section 3.4

- In inherited diseases, a defective gene or genes are passed on from parent to offspring. These diseases can be classified into two main groups, single-gene disorders and multifactorial disorders.

- Single-gene disorders can be recessive (xeroderma pigmentosum) or dominant (Huntington's disease and Marfan syndrome).

- Xeroderma pigmentosum (XP) is caused by the inheritance of a defective gene coding for one of the proteins involved in the repair of damaged DNA. People with XP display a range of symptoms dependent on the type and location of the mutation within the nucleotide excision repair mechanism genes.

- Huntington's disease is a neurodegenerative condition that develops in late middle age. Those affected have a greatly extended trinucleotide repeat sequence within a specific location in the huntingtin gene (*Hd*). The length of this sequence correlates with the severity of Huntington's disease symptoms experienced.

- Marfan syndrome is characterised by defective fibrillin, a major component of the microfibrils making up the matrix of connective tissue.

- Both Marfan syndrome and Huntington's disease show variable expressivity.

3.5 Neurodegeneration

In this chapter so far, you have met with diseases that result from infective agents that cause cell death or damage and also inherited conditions caused by gene mutations leading to the expression of disease phenotypes. In this section, you will investigate a further example of a neurodegenerative condition – **Parkinson's disease**, which is linked to the loss of certain types of brain cells. You have already encountered one neurodegenerative disease, Huntington's disease. The symptoms of Huntington's are caused by death of neurons in the brain, but in that case the loss of cells has a very clearly identifiable genetic link. In Parkinson's disease the genetic link is much weaker. The majority of people with Parkinson's have no close relations who have the disease; however, a very small proportion of cases (about 5%) do appear to have an inherited element. Parkinson's is thought to be a multifactorial disease deriving from multiple environmental factors, in some cases acting on genetically predisposed individuals as they age. Yet the signs of Parkinson's at the cellular level show similarities with other neurodegenerative diseases like Huntington's disease, a single-gene disorder, and Alzheimer's disease, a multifactorial disease.

Parkinson's disease without an identifiable genetic cause is referred to as *idiopathic* Parkinson's. A characteristic that both Huntington's and Parkinson's diseases share is that they are **proteinopathies** – diseases where structurally abnormal proteins within cells can lead to impaired functioning and frequently cell death.

- Both Parkinson's and Huntington's are diseases of middle to old age; suggest why the symptoms of the disease are not manifest in younger people.

- There may be a number of different reasons. It may take time for sufficient concentrations of the rogue proteins to build up within vulnerable cells; a gradual accumulation throughout life will ultimately lead to a point where cells are affected or die. In addition, a large proportion of vulnerable cells need to be lost before there is noticeable impact on the body. Alternatively, it may take prolonged exposure to environmental causative agents to result in disease symptoms.

3.5.1 Case study 4: Parkinson's disease

Parkinson's is one of the two most prevalent neurodegenerative diseases to afflict the human population (the most prevalent is Alzheimer's disease, another proteinopathy, again a disease inextricably linked to ageing and which is also quite poorly understood). In the UK, Parkinson's affects 100–180 people per 100 000 of the population. Approximately 0.4% of the population over 40 years of age have this disease, rising to 1% of the over 60s and 3–4% of people over 85. Around 10% of the institutionalised elderly have Parkinson's, although the UK government believes that as many as 40% of cases of Parkinson's are undiagnosed. Worldwide the estimated number of people with Parkinson's is over 6.3 million. This is truly a devastating disease and the complexity and range of symptoms shown, the rapid decline of patients and the difficulty in attributing a clear underlying cause, have impeded the progress of scientists who are working towards therapies and treatments.

Symptoms of Parkinson's disease

The onset symptoms of Parkinson's disease frequently include a 'resting tremor', usually in the hand. Once the individual initiates movement, the tremor goes away – hence the term resting tremor. About 70% of those with the disease usually show this symptom. The tremor will become more severe with time and spread up the limb and also affect the foot on the same side of the body. Ultimately, the tremor can spread to both sides of the body.

Another common symptom is bradykinesia or slowness of movement. People with Parkinson's can find it hard to initiate movement, they walk slowly with a shuffling gait and the facial muscles can lose their ability to control expression. Often the facial expression is said to be mask-like; completely unable to register emotions.

Finally, Parkinson's affects the ability to move muscles in the whole body leading to a rigidity and inflexibility. Ultimately, cognitive ability is impaired by dementia (a loss of cognitive ability beyond that attributable to normal ageing), a common and very distressing symptom.

A collection of neurons in the substantia nigra (a region of the basal ganglia in the midbrain) are directly concerned with the regulation and coordination of muscle contraction and therefore of voluntary movement. The root cause of the Parkinson's symptoms is related to depletion of a neurotransmitter called **dopamine** in this area of the brain. Many scientists attribute this lack of dopamine to the death of large numbers of dopamine-producing (dopaminergic) neurons in the substantia nigra, whilst others believe that the cells are not lost but rather inactive. Recent studies on the small proportion of Parkinson's sufferers whose disease seems to have a genetic rather than sporadic cause, do indeed suggest that the situation is much more complex then simple death of neurons. There is some evidence that synaptic dysfunction not cell death, is the heart of neurodegeneration. Certainly, replacing lost dopamine temporarily restores normal movement and reduces tremor in individuals whose Parkinson's is not very advanced. The most successful early treatment is with the drug Levodopa, a dopamine precursor that is converted in the brain to active dopamine. The belief that this condition may be a direct result of cell death in dopaminergic cells has stimulated interest in the use of stem cell technology as a potential long-term therapy for Parkinson's. The premise is that the deficiency in dopamine producing cells could be made good by inserting cells that are capable of synthesising dopamine. You will investigate the potential of stem cells to do just this in Chapter 4.

Parkinson's is linked to two characteristic changes in brain structure, namely a loss of dopaminergic neurons and the presence of **Lewy bodies** in the remaining dopaminergic cells of the substantia nigra. Lewy bodies are inclusions within the cytoplasm of the neurons, consisting of large aggregates of the protein alpha-synuclein (Figure 3.23). This is a short (140 amino acid) protein expressed in presynaptic terminals of vertebrate neurons. It is thought to regulate the transport of dopamine and its release from presynaptic vesicles at the synapse.

The synapse is an area of close contact between adjacent neurons.

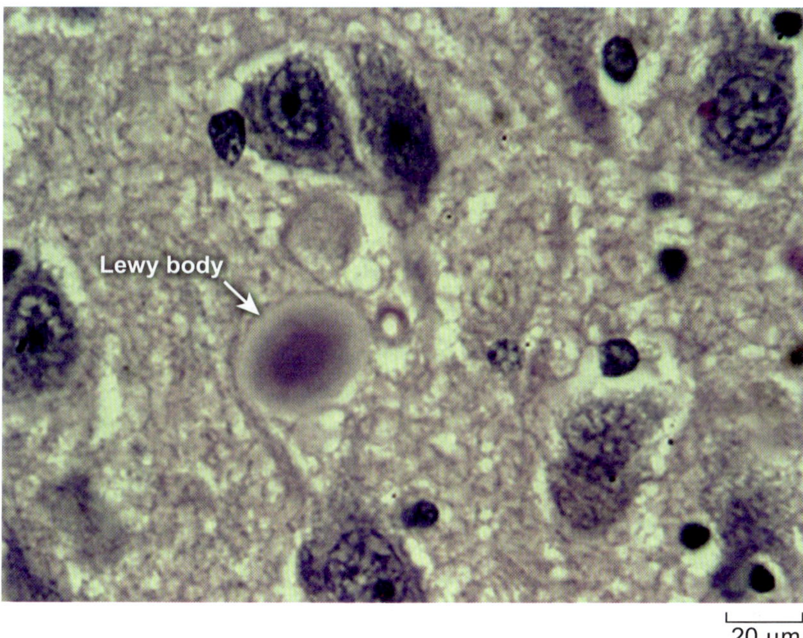

Figure 3.23 Micrograph of brain cells containing a Lewy body, which is an abnormal aggregation of protein.

Recent evidence indicates that accumulation of these protein aggregates in Lewy bodies plays a part in the degeneration of neurons. The alpha-synuclein within the Lewy body is usually modified in some way, for example, it may be bound to ubiquitin, the short regulatory protein that tags protein in the cell for destruction by the proteasome (Book 2, Section 1.5.4).

■   What is the function of the proteasome?

☐   The action of the proteasome breaks down and removes unwanted protein molecules in the cell by disrupting the peptide bonds between the amino acid residues.

The presence of these large aggregates of insoluble ubiquitin-bound protein suggests that the action of the proteasome is inhibited in these affected neurons, so the protein, though tagged for destruction, is not being dealt with by the proteasome. But why is the cell synthesising so much alpha-synuclein? One suggestion is that the protein is misfolded and therefore dysfunctional (Book 2, Section 1.5.5). Indeed there are examples of familial forms of Parkinson's disease where closely related members of the same family have been found to carry a mutation in the alpha-synuclein gene. One such mutation produces misfolded alpha-synuclein and another overexpresses the alpha-synuclein gene. Both mutations lead to all the symptoms of Parkinson's described above. In other patients with idiopathic disease it seems that risk factors such as environmental exposure to neurotoxic chemicals can cause the same aggregation of this protein and the same effect on the neurons of the substantia nigra and consequent reduction of dopamine provision in the brain.

However, the role of protein aggregates within neurons in this disease may be only one piece of an elaborate jigsaw. Other potential causal factors for Parkinson's disease are being explored including the role of mitochondrial dysfunction within neurons (as in Huntington's), whereby much higher levels of toxic oxidative molecules are produced from the electron transport chain (Book 2, Section 3.6.3). This could increase damage within the mitochondria, leading to cell death. There seem to be many interrelated causes for Parkinson's, some resulting from genetic mutation – 11 different genes have now been found to be significant in Parkinson's disease – and some from environmental stresses. Possession of any of the gene mutations increases the risk of developing Parkinson's but individuals without any of these genetic risk factors can also contract the disease. Exposure to pesticides such as Paraquat (weed killer) and environmental pollutants such as mercury, aluminium and copper are known to be risk factors for Parkinson's. Similarly lifestyle related issues, such as lack of certain vitamins, high dietary iron and a high meat intake, have also been implicated. Unlike Huntington's disease, where a single mutated gene is involved, in Parkinson's disease we are facing a disease where complex interactions between genes and environment determine susceptibility. Only with greater understanding of all the competing risk factors will new therapies and treatments emerge for this most distressing of age-related illnesses.

Now you are aware of the complexity of the causal factors for neurodegenerative diseases such as Parkinson's disease, complete Activity 3.5 to find out more about the cellular basis for this and other neurodegenerative diseases.

Activity 3.5 Neurodegenerative conditions

(LOs 3.1 and 3.9) Allow 120 minutes to access and view the online resources

In this activity you will access a number of up-to-date online resources to allow you to investigate aspects of the cellular basis of three neurodegenerative diseases (Parkinson's, Alzheimer's and Huntington's) and how changes at the cellular level relate to the pathology and symptoms of these diseases.

Summary of Section 3.5

* Parkinson's is a common neurodegenerative disease.
* The signs of Parkinson's at the cellular level show some similarities with other neurodegenerative diseases like Huntington's disease, a single-gene disorder.
* Parkinson's is linked to changes in brain structure, namely a loss of dopaminergic neurons.
* Parkinson's is a disease where complex interactions between genes and environment determine susceptibility.

3.6 Final word

In this chapter you have briefly studied a broad range of human and animal diseases with a variety of different causes: bacterial, viral, genetic or even a combination of factors. In each case you have concentrated on the effect of the disease at the cellular level before delving into the effects these diseases have on individuals and communities.

In Chapter 4 you will conclude your studies in cell biology with an examination of some new ways of exploiting our understanding of this exciting field – so-called cell technologies. This will include an examination of the development of novel antimicrobials that aim to alleviate the risks posed by infectious diseases, and a look at recent advances in stem cell technology, including a possible therapy for the treatment of neurodegeneration. You will also look at the impact that genetic engineering has on the production of highly valuable therapeutic substances such as insulin.

3.7 Learning outcomes

3.1 Explain some of the underlying causes of disease (infectious agents, inherited disease, degenerative disease) at the cellular level.

3.2 Describe the impact of infectious agents such as viruses and bacteria on their hosts.

3.3 Explain, in simple terms, how the innate and adaptive immune systems are able to combat infectious disease.

3.4 Outline a range of bacterial virulence factors and explain how possession of these characteristics contributes to enhanced bacterial pathogenicity.

3.5 Outline methods of controlling epizootics of FMD and compare and contrast the methods used.

3.6 Explain in terms of the underlying mutation, the clinical phenotypes of individuals with xeroderma pigmentosum, Marfan syndrome and Huntington's disease.

3.7 Demonstrate how these diseases are inherited within families by interpretation of simple human pedigrees.

3.8 Explain the term 'variable expressivity' with reference to Marfan syndrome and Huntington's disease.

3.9 Describe and relate different types of neurodegenerative diseases (principally Parkinson's disease, Huntington's disease and Alzheimer's disease) to their underlying causes.

References

Song, W., Chen, J., Petrilli, A., Liot, G., Klinglmayr, E., Zhou, Y. et al. (2011) 'Mutant huntingtin binds the mitochondrial fission GTPase dynamin-related protein-1 and increases its enzymatic activity', *Nature Medicine*, vol. 17, pp.377–82; also available online at http://www.nature.com/nm/journal/v17/n3/full/nm.2313.html

Chapter 4 Cell technologies

4.1 Introduction

In this module so far, you have studied many aspects of cell biology, including how cells reproduce, control the expression of their genes, obtain energy and signal to surrounding cells. This wealth of knowledge about cells holds great potential for real-life applications, including the modelling and control of disease and the synthesis of new drugs and other useful products from cells. Indeed you began the module with a demonstration of cell technology in action: Activity 1.1 in Book 1 demonstrated how biodiesel could be efficiently manufactured by genetically modified organisms using simple sugars as a nutrient source.

The harnessing of the properties of cells is not a new phenomenon: humans have been exploiting their biological processes for millennia. Baking, brewing and winemaking have traditionally relied on yeast fermentation to manufacture products on both a domestic and an industrial scale. Modern cell technology is used to enhance or optimise such natural processes within living organisms to maximise production. Indeed, many of the industrial processes used in the biotechnology industries are based on well-understood fermentation technologies from the traditional industries. Newer applications of biotechnology include the harnessing of cells naturally capable of synthesising substances such as antibiotics, but also the use of genetic engineering to modify cells in order that they can synthesise products they would normally be unable to make. Hormones, synthetic antibiotics, antibodies, enzymes, biofuels and a variety of biopolymers have been produced using cultured recombinant cells. More recently, the focus has shifted to the characteristics of the cell itself, rather than its products. The culture of animal cells with particular properties (e.g. specific types of human stem cells) is one of the most recent and exciting biotechnological challenges.

In this chapter you will learn about three areas of cell technology that bring the promise of new therapies to combat a range of human diseases. You will begin in Topic 1 with the development of new antibacterial compounds that may help to combat increased resistance to current antibiotics. Topic 2 will focus on the impact of recombinant DNA technology and the capabilities it brings to the manufacture of human proteins on an industrial scale. Finally, in Topic 3, you will focus on the enormous potential of stem cells and the promise of cell replacement therapy to control human disease.

The main concepts underpinning each of the three topics will be introduced in this chapter, and you may also wish to return to earlier chapters and revisit many of the fundamental aspects of cell biology previously touched on in the module. At the end of each topic, you will complete an activity that will further develop your understanding of these concepts and allow you to access current developments via the module website.

4.2 Topic 1: The end of the antibiotic age?

In Chapter 3 you were introduced to the concept of bacterial virulence factors, including antibiotic resistance. You now have the opportunity to delve a little further into the role of antibiotics in fighting infection, from the structure of antibiotics to their synthesis and mechanisms of action against different classes of bacteria. You will then look at various molecular strategies evident in antibiotic-resistant bacteria and the genetic mechanisms that allow this resistance to spread so rapidly. Finally, you will investigate some new and alternative antibacterial techniques and therapies.

When introducing antibiotics to medical students in 1968, the renowned bacteriologist L. P. Garrod wrote:

> No one recently qualified, even with the liveliest imagination, can picture the ravages of bacterial infection which continued until little more than 30 years ago…. Pneumonia was a common cause of death even in young and vigorous patients, and puerperal fever (post partum septicaemia) had a high mortality, little affected by any treatment then available.
>
> *Garrod and O'Grady (1968)*

Post partum septicaemia is blood poisoning in the mother following childbirth.

The widespread introduction of antibacterial drugs into clinical practice began in the late 1930s, and revolutionised medicine at that time (Figure 4.1) but we have to ask ourselves: for how much longer will antibiotics be effective?

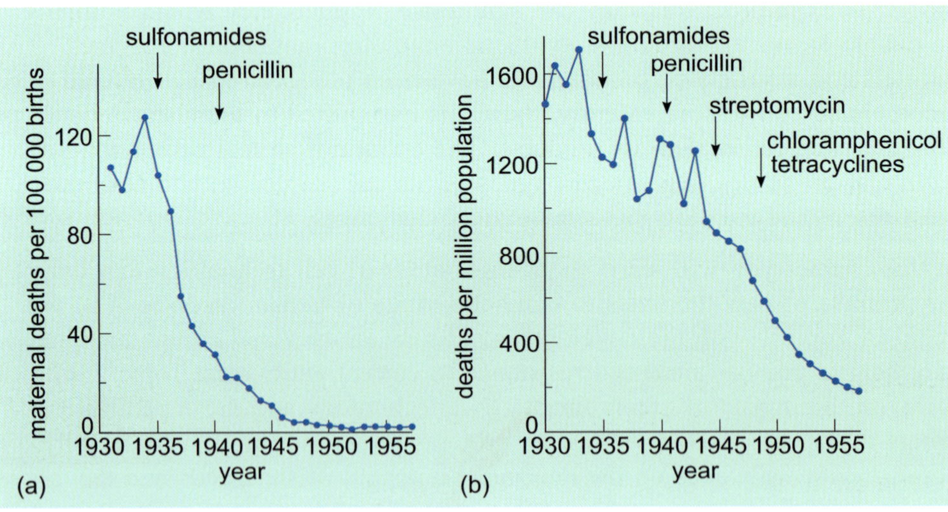

Figure 4.1 Effect of antibiotics on death rates in England and Wales between 1931 and 1957 from (a) childbirth-related infection (puerperal fever) and (b) all infectious disease. Data from Barber (1960). The arrows indicate the time of the introduction of specific antibiotics.

4.2.1 Antibiotics defined

Antibiotics can be defined as chemicals that kill or inhibit the growth of microbes (Figure 4.2).

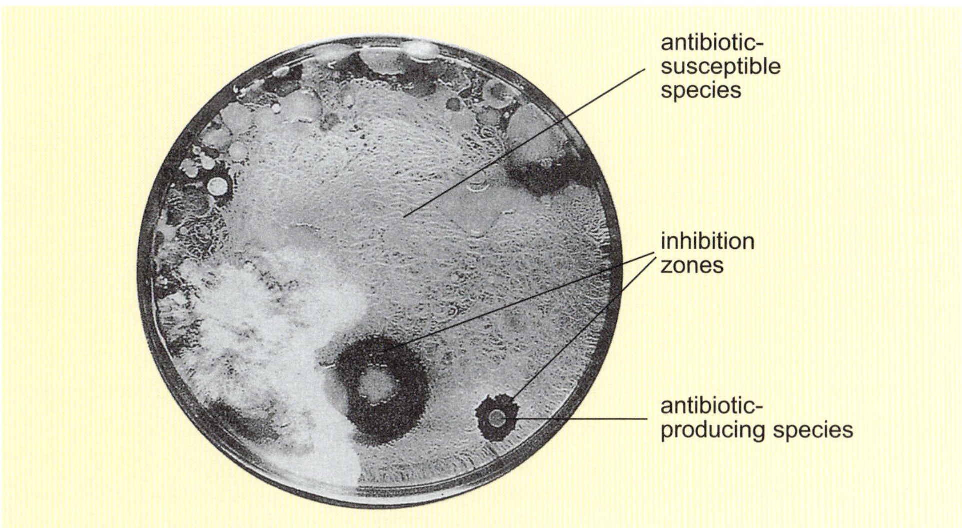

Figure 4.2 The antibiotic action of soil bacteria growing on an agar plate. The small round colonies surrounded by clear zones are *Streptomyces* species. The large irregular colonies are *Bacillus* species whose growth is inhibited by antibiotic secretions from *Streptomyces*.

Many of the antibiotics in clinical use are based on naturally occurring ones, synthesised by microbes such as bacteria or fungi, but usually chemically modified to improve performance. Others are completely synthetic. Comparatively few species secrete antibiotics; examples are the fungus *Penicillium* and various species of the bacterial genera *Bacillus* and *Streptomyces*. Most of these compounds are antibacterial agents, but the increase in AIDS, other viral diseases and fungal diseases has stimulated the search for antiviral and antifungal compounds too. Several classes of drugs are considered here, although the emphasis is on antibacterials, since they have the widest application in medicine and agriculture, and are particularly lucrative for the pharmaceutical industry.

Bacteria are said to be either sensitive or resistant to a particular antibiotic. Resistant bacteria continue to grow, even in the presence of high concentrations of antibiotic, while sensitive bacteria are unable to do so even at low antibiotic concentrations. Resistance to a certain antibiotic can be easily demonstrated by growing the bacterial culture on an agar plate and placing paper discs or strips impregnated with antibiotic on the agar (Figure 4.3). As the 'lawn' of bacteria grows, clear inhibition zones appear wherever an effective dose of antibiotic is encountered.

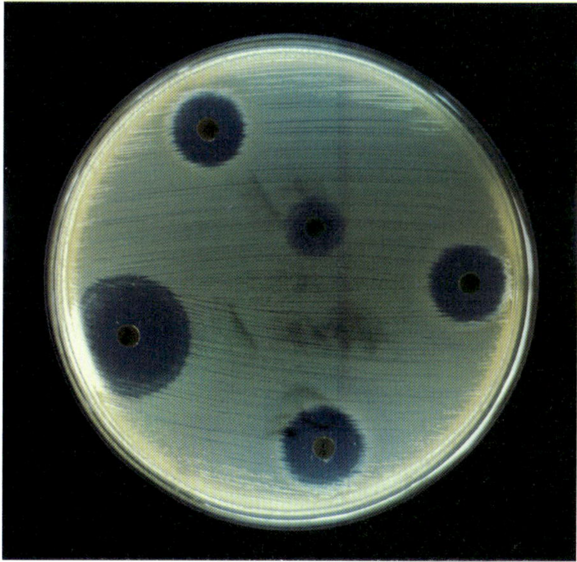

Figure 4.3 A lawn of *Staphylococcus albus*. Discs impregnated with varying concentrations of antibiotic are placed on the agar and the plate incubated for 24 hours.

4.2.2 Synthesis of antibiotics

The first recorded antibacterial agent used in medicine was an entirely synthetic chemical; an arsenic-based anti-syphilis drug developed by Paul

Ehrlich in the early 1900s. Later, the sulfonamides (also entirely synthetic) were used to treat a range of bacterial infections. Then, in 1939, a group of Oxford scientists, motivated by the impending war, scaled up Alexander Fleming's work on penicillin, the active ingredient in bactericidal secretions from the fungus *Penicillium*. Penicillin had been discovered purely by accident in 1928, when Fleming, who at the time was studying *Staphylococcus*, noticed clear, bacteria-free zones surrounding the contaminating colonies of *Penicillium* fungus that had been blown on to his supposedly sterile culture plates.

Today, antibiotics have enormous commercial importance; well over 100 000 tonnes are sold each year worldwide. The majority are now synthetic rather than naturally derived. Indeed, surprisingly few genera are used commercially in antibiotic production, the most notable being certain filamentous fungi and the actinomycete bacteria. Antibiotics are complex chemicals; microbes synthesise them stepwise in a long series of enzyme-catalysed reactions. The starting compounds are usually products of the cell's housekeeping metabolic pathways, for example, amino acids, purine and pyrimidine bases and acetyl CoA. These essential compounds are synthesised in the exponential phase of growth (Section 1.2.1), while secondary products such as antibiotics are made only in the stationary phase, presumably when resources can be diverted away from growth. In commercial processes, pure cultures of antibiotic-producing bacteria and fungi are grown in huge bioreactors (containing thousands of litres). The chemically labile antibiotic products are then harvested and purified. You will look at some of the process technology required to grow bacteria and fungi on an industrial scale in Topic 2 (Section 4.3).

4.2.3 Classifying antibiotics

Each class of antibiotics has characteristic structural features that define the specificity for target molecules in the bacteria they affect. In this respect, they are like enzyme inhibitors, and indeed many of the targets are enzymes, or other proteins.

Antibiotics can be classified into broad- or narrow-spectrum types. The so-called broad-spectrum antibiotics, which affect both Gram-positive and Gram-negative bacteria, are understandably popular, particularly where the underlying pathogen cannot readily be identified. However, overuse of broad-spectrum antibiotics is widely considered to have led to widespread resistance, and there is increasing pressure to use narrow-spectrum drugs where possible. Some other factors that may influence choice of drug are included in Table 4.1, for example, toxicity, and whether the drug can be taken orally. The oral route is obviously preferable to injection or infusion, where trained health workers have to administer the drugs.

The most accessible target is the peptidoglycan of the bacterial cell wall that lies outside the cell membrane. It is a thick layer in Gram-positive bacteria and considerably thinner in Gram-negatives (Book 1, Section 3.3). However, Gram-negative bacteria also have another layer, the outer membrane (containing lipopolysaccharides and lipoproteins), which water-soluble

Table 4.1 Some antibiotics in clinical use.

Class or name	Examples	Comments
penicillins (β-lactam type antibiotics)	benzyl penicillin, ampicillin, amoxicillin, methicillin, flucloxacillin	Although early penicillins were directed against Gram-positive bacteria, modern ones are also effective against Gram-negative bacteria
cephalosporins (β-lactam type antibiotics)	cefalexin	Broad spectrum; can be taken orally
carbapenems (β-lactam type antibiotics)	imipenem	Broad spectrum; useful for nosocomial (hospital-acquired) infections
fluoroquinolones	ciprofloxacin	Active against Gram-negatives (especially urinary tract infections) and some Gram-positives; increasing resistance reported; can be taken orally
tetracyclines	oxytetracycline	Broad spectrum; often used for acne; use of this class of antibiotics is severely limited by resistance
chloramphenicol		Broad spectrum; often used for eye infections
macrolides	erythromycin	Active only against Gram-positives (e.g. *Legionella*); useful for patients allergic to penicillin
aminoglycosides	Streptomycin, gentamycin	Formerly used to treat tuberculosis but now limited by resistance; can be taken orally
sulfonamides	sulfadiazine	Broad spectrum; limited by resistance; entirely synthetic
glycopeptides	vancomycin	Active only against Gram-positives; useful for nosocomial (hospital acquired) infections, but resistance is increasing; can be toxic to inner ear and kidney
rifampicins	Rifampicin A	Formerly used for tuberculosis but resistance is increasing
lipopeptides	Daptomycin	Active only against Gram-positives; useful against multiresistant bacteria

molecules can cross only via specific membrane protein complexes called porins (Figure 4.4). This outer membrane, for example, excludes penicillin, which is thus ineffective on most Gram-negative bacteria. Antibiotics with an intracellular target must cross these outer layers before penetrating the cell membrane by simple diffusion or, in a minority of cases, by mimicking a compound naturally taken up by the cell and 'hitching a ride' on a transport protein to enter the cell by facilitated diffusion or active transport (Book 2, Section 2.8).

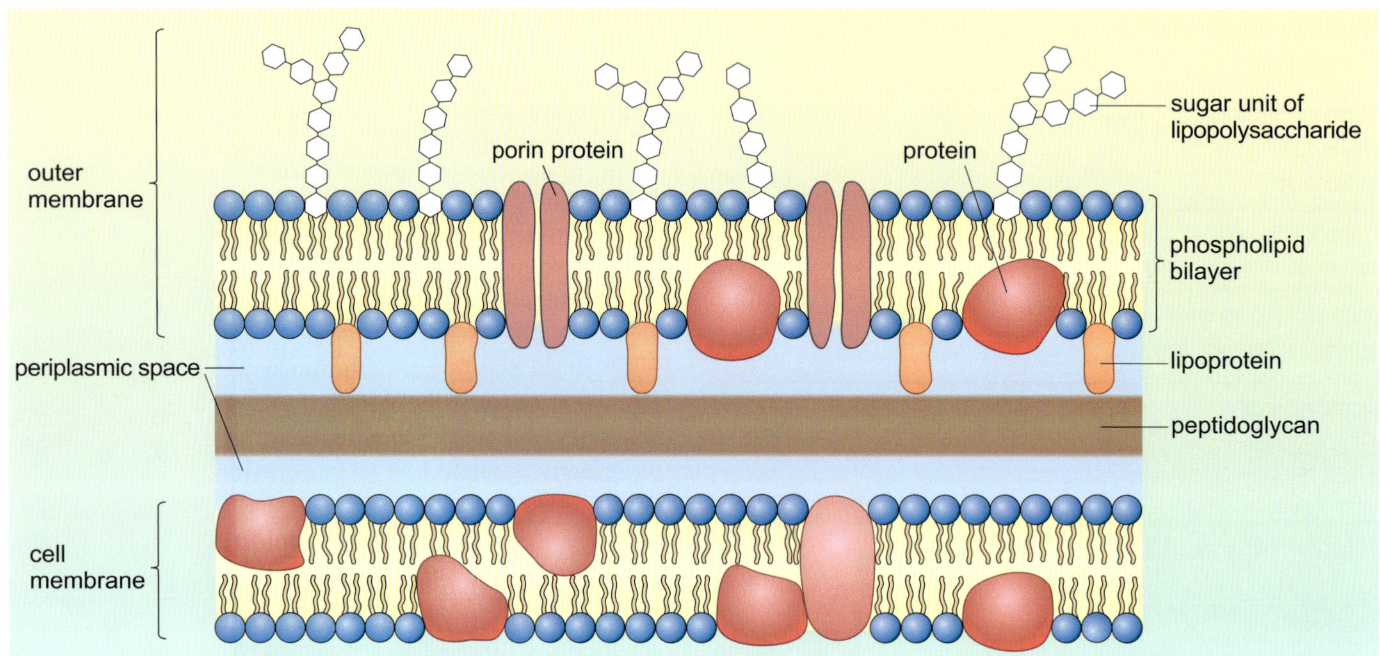

Figure 4.4 Diagram showing the composition of the cell membranes and cell wall of a Gram-negative bacterium.

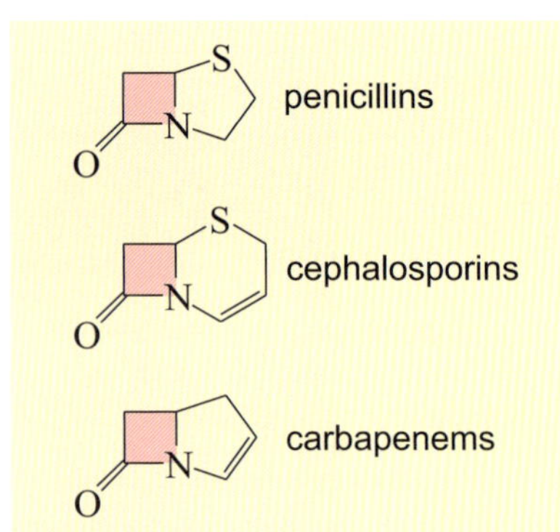

Figure 4.5 Core ring structures of three classes of β-lactam antibiotics. The β-lactam ring is shaded pink in each case.

The β-lactam antibiotics (based on penicillin) are extremely useful, comprising over half the clinical market for antibacterial drugs. As discussed in Section 3.2.3, they act by inhibiting the synthesis of peptidoglycan. All of them have the same four-membered cyclic amide ring, called a β-lactam ring (Figure 4.5). Penicillins, cephalosporins and carbapenems differ only in the structure of the ring to which the β-lactam ring is joined. Only a few of the penicillins attack Gram-negative bacteria, while cephalosporins and carbapenems are, in general, more broad-spectrum antibiotics. Resistance to penicillins is increasingly limiting their effectiveness (see Section 4.2.5), and it is feared that without vigorous controls the same may happen with all the β-lactam antibiotics.

Structures of some non-β-lactam drugs are shown in Figure 4.6. Among the older broad-spectrum antibiotics are sulfonamides, tetracyclines and chloramphenicol (Table 4.1), all of which, in different ways, disrupt the synthesis of bacterial nucleic acids or proteins.

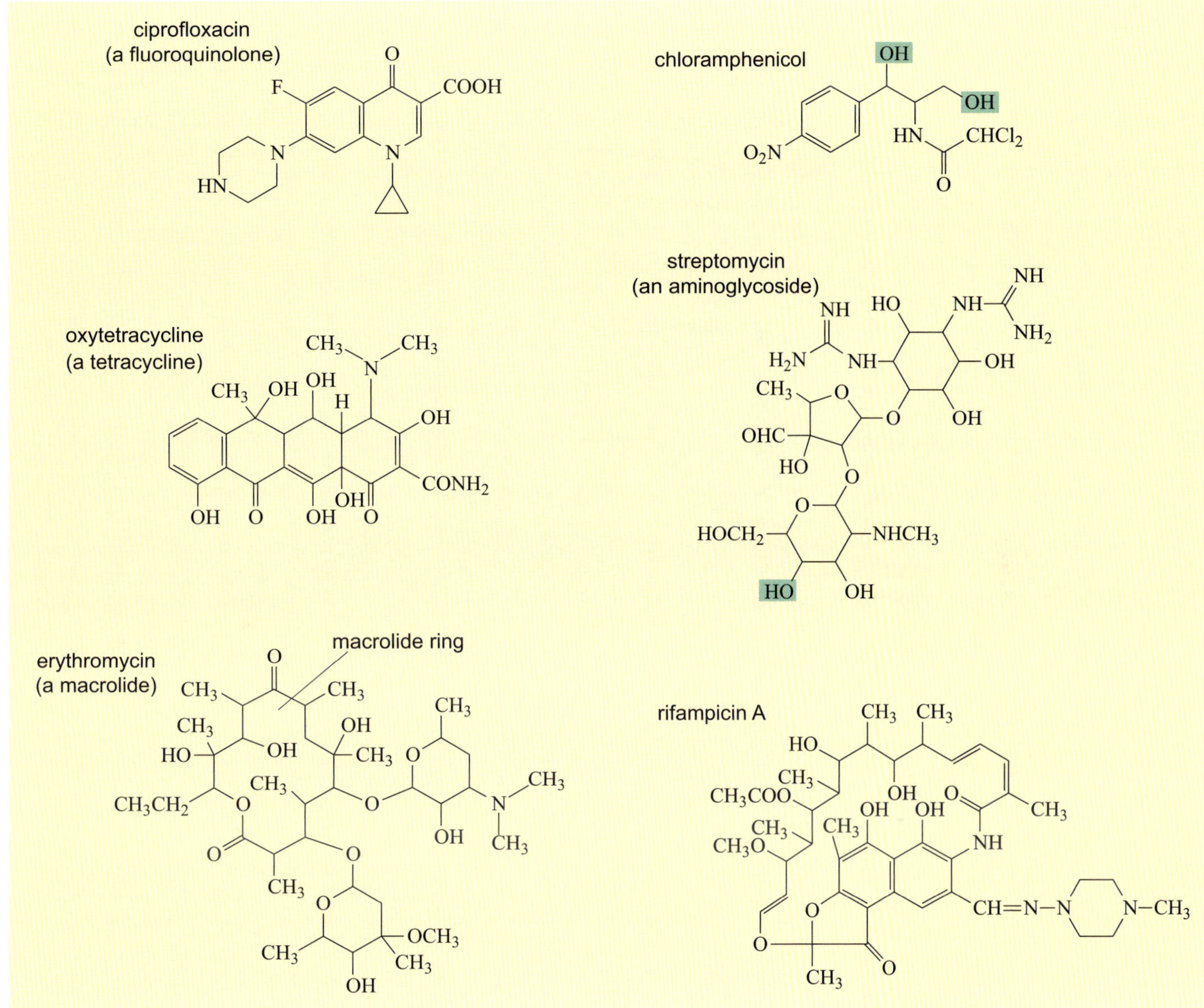

Figure 4.6 Structure of some non-β-lactam antibiotics. The green boxes on chloramphenicol and streptomycin indicate the target sites of specific inactivating enzymes that produce antibiotic resistance.

4.2.4 Selective toxicity

According to the principle of selective toxicity, the ideal antibiotic damages the bacterial pathogen without harming its human host. This depends on finding bacterial targets that do not exist in the human host, or are sufficiently different at the molecular level that they are not affected by antibacterial drugs. Some of the molecular targets of antibacterials are summarised in Figure 4.7. Particularly useful targets include the cell wall and nucleic acid and protein synthesis.

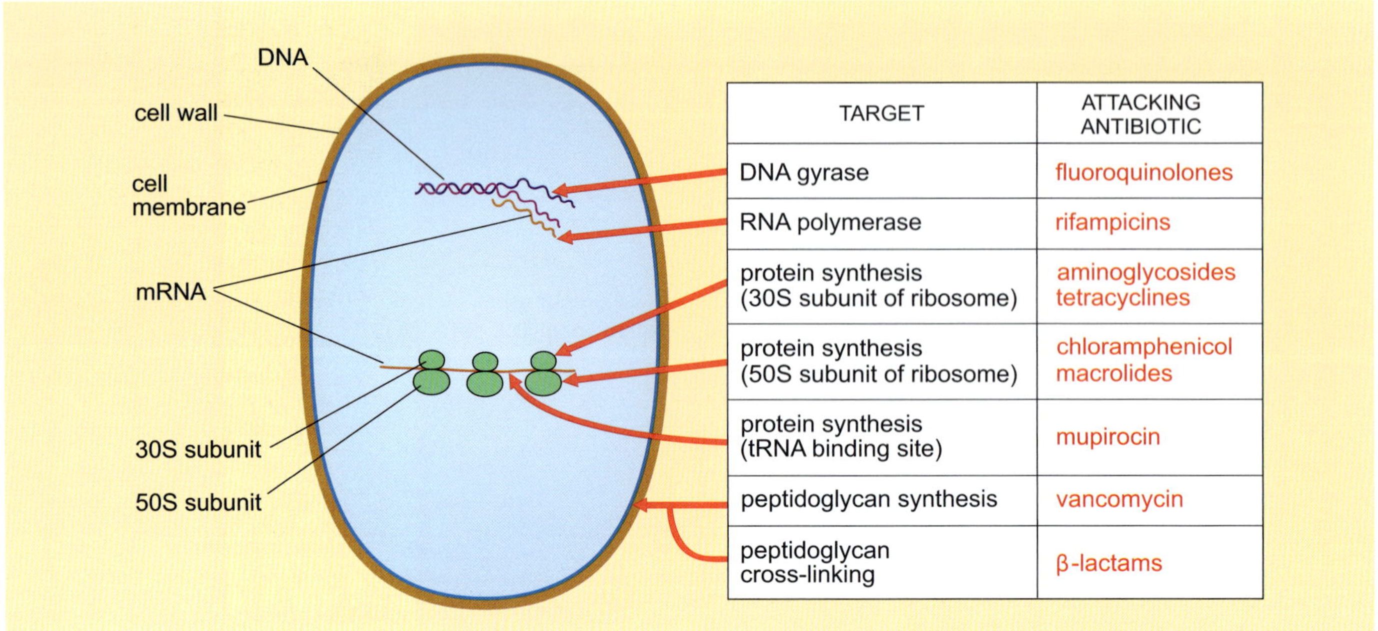

TARGET	ATTACKING ANTIBIOTIC
DNA gyrase	fluoroquinolones
RNA polymerase	rifampicins
protein synthesis (30S subunit of ribosome)	aminoglycosides tetracyclines
protein synthesis (50S subunit of ribosome)	chloramphenicol macrolides
protein synthesis (tRNA binding site)	mupirocin
peptidoglycan synthesis	vancomycin
peptidoglycan cross-linking	β-lactams

Figure 4.7 Target sites for antibacterial agents.

Targeting the cell wall

The bacterial cell wall in part comprises a rigid layer of cross-linked peptidoglycan chains which prevents the underlying cell membrane from lysing (bursting) as the bacteria absorbs water by osmosis (Figure 4.8). So anything that interferes with cell wall synthesis is particularly toxic to growing cells. The structure of the β-lactam ring (Figure 4.5) is very similar to amino acid residues on the subunits used to synthesise peptidoglycan and the β-lactam antibiotics bind irreversibly to the active site of transpeptidase, the enzyme that catalyses peptidoglycan cross-linking. The β-lactams are much used in medicine because they are not toxic to humans and other animals, since the cells of these hosts have nothing equivalent to the bacterial cell wall. Vancomycin also interferes with the synthesis of peptidoglycan, blocking an enzyme earlier in the sequence, but it is active against only Gram-positive bacteria as the molecule is too large to cross the Gram-negative outer membrane.

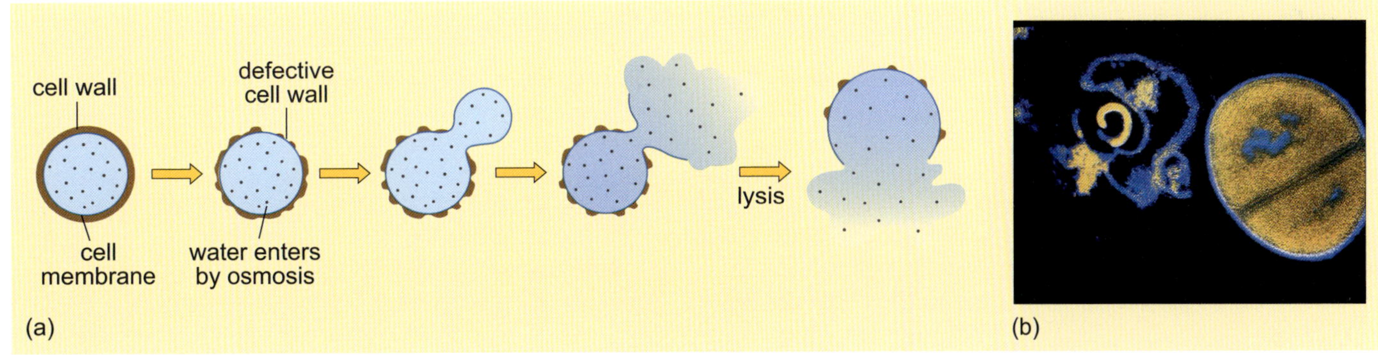

Figure 4.8 Lysis of a bacterium with a defective cell wall. (a) Diagram showing the sequence of events that lead to lysis. (b) Light micrograph of *Staphylococcus aureus*: a lysed cell is visible on the left and an intact dividing cell on the right.

Targeting nucleic acid and protein synthesis

Selective toxicity is rather more difficult to achieve when the synthesis of nucleic acids and proteins is the target.

■ Suggest why selective toxicity is difficult to achieve in this case.

□ Because animals as well as bacteria synthesise nucleic acids and proteins.

However, there are differences between the proteins that carry out these processes in bacteria and eukaryotes. Fluoroquinolones are extremely useful antibiotics that specifically inhibit the bacterial topoisomerase enzymes that relieve torsional strain during unwinding of the DNA helix (Book 1, Section 5.3).

■ Why would inhibiting these enzymes block nucleic acid replication in bacteria?

□ Without the activity of topoisomerases, DNA helicase could not continue to separate the DNA strands and DNA replication would cease.

Rifampicin, once an important antibiotic in the treatment of tuberculosis, exerts its antibacterial activity by binding specifically to the β and β′ subunits of bacterial RNA polymerase (Book 1, Section 6.3.1).

■ What process does rifampicin inhibit?

□ Transcription of bacterial DNA.

The aminoglycoside antibiotics streptomycin and gentamycin (Table 4.1) bind specifically to the 30S subunit of bacterial ribosomes inhibiting the translocation of tRNA during translation of mRNA to protein. Tetracyclines inhibit the binding of tRNA to bacterial ribosomes, while chloramphenicol acts on the 50S subunit to block elongation of the polypeptide (Book 1, Section 6.6.3).

4.2.5 Resistance to antibacterial drugs

There are four different molecular mechanisms that confer antibiotic resistance in bacteria.

Antibiotic-inactivating enzymes

Perhaps the most notorious antibiotic-inactivating enzymes are the β-lactamases, which, as the name suggests, hydrolyse the β-lactam ring in penicillin-like antibiotics, preventing them from binding to the target site (Figure 4.9). Resistant Gram-positive bacteria secrete β-lactamase into their surroundings, while Gram-negative bacteria retain the enzyme within the periplasmic space near the target site of the antibiotic (the cell-wall-synthesising machinery). Even newer β-lactam drugs like the carbapenems (Figure 4.5) are now susceptible to hydrolysis, as bacterial strains evolve that can accommodate the new drugs into the active site of their β-lactamase. One successful clinical strategy is to use combination drugs such as Augmentin, which contains a β-lactam antibiotic and a β-lactamase inhibitor to block the active site of the β-lactamase enzyme so that it can't hydrolyse the antibiotic.

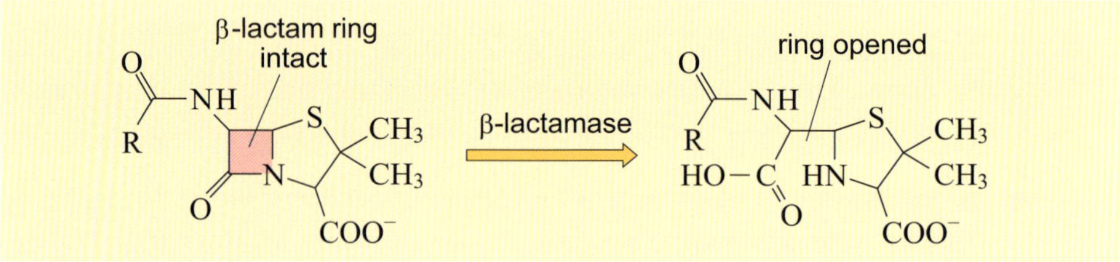

Figure 4.9 The β-lactamase-catalysed hydrolysis of a penicillin results in opening of the β-lactam ring and thus inactivation of the antibiotic.

Other antibiotic-modifying enzymes confer resistance to aminoglycosides, e.g. streptomycin, by adding bulky phosphate or other groups to reactive amino ($-NH_2$) or hydroxyl ($-OH$) groups on the antibiotic (Figure 4.6). Thus modified, the antibiotic no longer binds to its target site on the 30S ribosomal subunit (Figure 4.7). Numerous other antibiotic-modifying enzymes have evolved, each able to inactivate specific antibiotics.

Reduction in permeability of the bacterial cell boundary

No antibiotic is effective unless it can reach its target site. Many bacteria become resistant to a particular antibiotic by reducing their permeability to it.

- Which of the following layers must be permeable to antibiotics targeted at nucleic acid or protein synthesis in Gram-negative bacteria: outer membrane, peptidoglycan layer and cell membrane (Figure 4.5)?

- All three layers must be crossed for the antibiotic to reach susceptible molecules (e.g. target enzymes, ribosomal proteins) in the interior of a Gram-negative bacterium.

The cell walls of Gram-positive bacteria are permeable to most antibiotics, but Gram-negative bacteria can become resistant by increasing the closeness of packing of the lipopolysaccharide molecules in the outer membrane. Alternatively, modification of the porin proteins that usually provide water-filled channels through this layer (Figure 4.4) can exclude polar (charged) antibiotics. The cell membrane is freely permeable to non-polar antibiotics, but polar antibiotics rely on membrane carrier proteins, which also have a transport role in the normal functioning of the bacterial cell. These proteins too can become modified (via mutation) to reduce their antibiotic affinity, thereby conferring resistance.

Active efflux of antibiotic molecules

Efflux pumps that expel antibiotics out of the cell against a concentration gradient (i.e. by active transport) have been found in the cell membrane of Gram-positive bacteria, and in the outer membrane of Gram-negatives. Such active transporters often have a very broad specificity, pumping out a wide variety of chemically unrelated compounds that are not required by the cell. They may have evolved from membrane transport proteins that predate the antibiotic era.

Changes in target molecules

This very effective mechanism is the most widespread of all. In rapidly proliferating bacteria, a gene that encodes an antibiotic target molecule can readily mutate to an altered form that can become more widespread in the population if it confers a growth advantage. The mutant target molecule may retain the activity of the original version, but no longer be affected by a specific antibiotic (Figure 4.10). Examples of this bacterial strategy include the cell-wall-synthesising enzymes inhibited by β-lactams. These enzymes may become modified, decreasing their antibiotic affinity but without impairing enzymatic activity.

Similarly, the components needed for nucleic acid or protein synthesis (Figure 4.7) may become modified so that they no longer bind antibiotics such as streptomycin (by changes to a 30S ribosomal subunit protein), fluoroquinolone (by changes to DNA topoisomerase), or rifampicin (by changes to the β and the β′ subunit of RNA polymerase).

In Section 3.2.3 you touched on how antibiotic resistance can arise and spread through bacterial populations when you studied the effect of *Staphylococcus aureus* on the cell. Recall that the resistant strain of this bacterium is called MRSA.

■ Which antibiotic is MRSA resistant to and how is this resistance acquired?

☐ Methicillin. MRSA has different penicillin binding proteins on its surface which cannot bind methicillin (possibly a result of horizontal transmission of the SCCmec transposon from other cells in the population or even from other bacterial species).

The rise of antibiotic resistance means that new antibiotics are constantly being identified, synthesised and brought to market. Most are modified versions of existing approved antibiotics, but there is an increasing demand

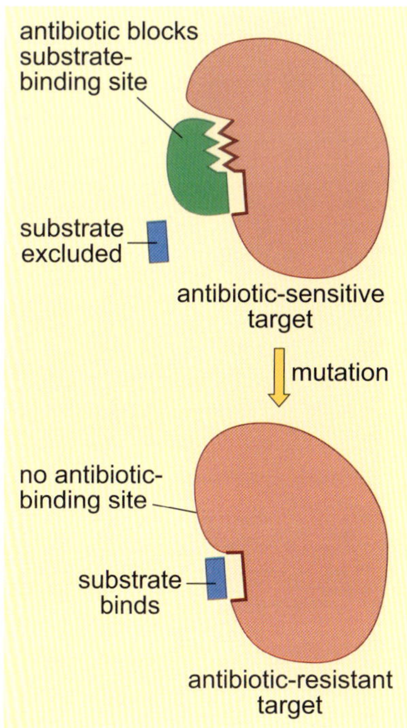

Figure 4.10 Schematic diagram showing how a structural change in a target protein as a result of mutation can produce antibiotic resistance.

for novel antibacterials with alternative modes of action. To complete this topic, you will look at a recently introduced antibiotic, Daptomycin, and also the possibility of using bacteriophage and quorum sensing behaviour as the basis for new antibacterial therapies. These three issues will be briefly introduced here and further developed in Activity 4.1.

4.2.6 Daptomycin: a new antibiotic for Gram-positives

Daptomycin (its trade name is Cubicin) is a cyclic lipopeptide antibiotic (Table 4.1), the first new class of antibiotics to be marketed in 30 years. Its activity against Gram-positive pathogens such as MRSA makes it of particular interest, given that in Europe around 25% of *S. aureus* infections are caused by MRSA and that use of vancomycin as a last resort treatment is now proving to be less effective. Indeed, at least three completely vancomycin-resistant *S. aureus* strains have now been isolated.

Daptomycin is produced by the soil bacterium *Streptomyces roseosporus*. Figure 4.11 shows its unique chemical structure, with a 13-membered amino acid ring and a lipophilic tail. Like many antibiotics, Daptomycin targets the cell membrane of prokaryotic cells but its chemical structure allows a unique mode of action. The lipophilic tail inserts itself into the cell membrane of Gram-positive bacteria; this causes the membrane to depolarise (i.e. undergo a change in membrane potential, Book 2, Section 2.8.3) as potassium ions are lost from the cell. This, in turn, results in cessation of protein and DNA synthesis and cell death. Unusually, Daptomycin is active against bacterial cells in the exponential and stationary phases of growth. Recall that the majority of bacterial cells in biofilms (Section 3.2.3) are in the stationary phase, and therefore normally very hard to kill.

Of course the real question regarding Daptomycin's effectiveness as a new therapy is how soon will the target pathogens become resistant. So far, Daptomycin seems to have low resistance potential. However, as a natural antibiotic produced by bacteria growing in the soil, it is likely that alleles conferring resistance to Daptomycin already exist, and it is only a matter of time before they are expressed in human pathogens.

Antibiotic production clearly relies on the large-scale cultivation of the producing bacteria or fungi. As you will see in Section 4.3, this requires suitable process technology to maximise the yield of target molecule and to purify it for use as a pharmaceutical. Antibiotics can be modified chemically after production, to change substituent groups. Such novel antibiotics can have improved therapeutic properties and may be better able to combat resistant strains of bacteria. Alternatively, genetic engineering of the source bacterium or fungus can alter their biosynthetic processes to make modified natural products from scratch. Gene technology has allowed, for example, the antibiotic erythromycin to be synthesised on an industrial scale by the rapidly growing *E. coli* rather than its natural producer, the soil bacterium *Saccharopolyspora erythraea*.

In addition to research into new antibiotics to combat developing resistance, scientists are trying to find other ways of targeting pathogens. The next two sections focus on two of these alternative methods: the use of bacterial viruses

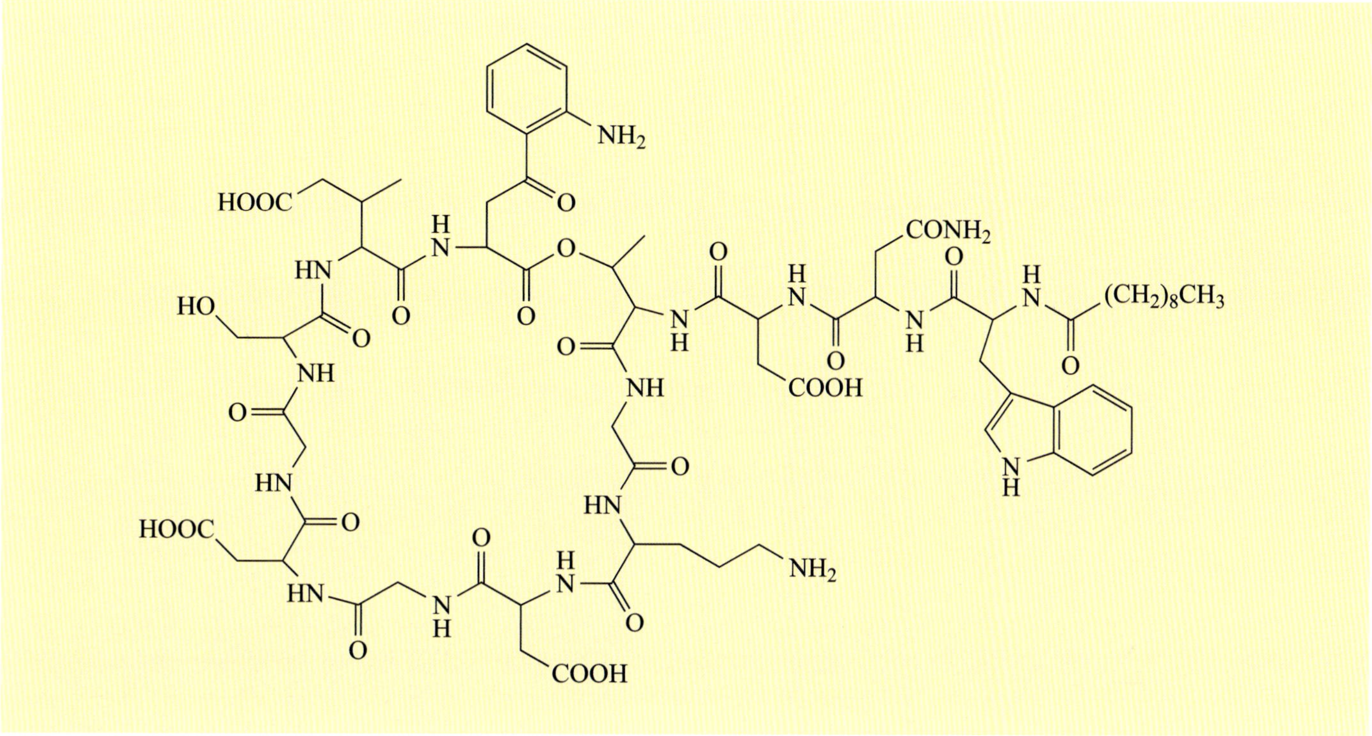

Figure 4.11 Chemical structure of Daptomycin showing the 13-membered amino acid ring structure on the left with the lipophilic tail projecting to the right of the structure.

(phage therapy) to destroy bacteria, and the potential for disrupting bacterial signalling (quorum signalling) to prevent bacteria from establishing an infection.

4.2.7 Phage therapy

Bacteriophages are the most abundant biological entities on Earth and estimates suggest that phage predation destroys half of the world population of bacteria every 48 hours. The use of phage therapy as a means of treating bacterial infections was first developed about 90 years ago by researchers in the former Soviet Union. In the 1990s, the increasing issues with antibiotic resistance led to a renewed interest in the use of phages. Phage therapy relies on the bactericidal nature of bacteriophages (Section 3.3.1); that is, their ability to lyse pathogenic bacteria without affecting the host. Once bacteria are lysed, the released virion progeny (between 10 and 100 virions per lysed cell) then continue the cycle, migrating to other sites of infection in the body (Figure 4.12). Phage therapy can be equally effective against antibiotic-sensitive and resistant bacteria, provided that a phage capable of infecting a given bacterial strain is available – recall that bacteriophages are highly specific.

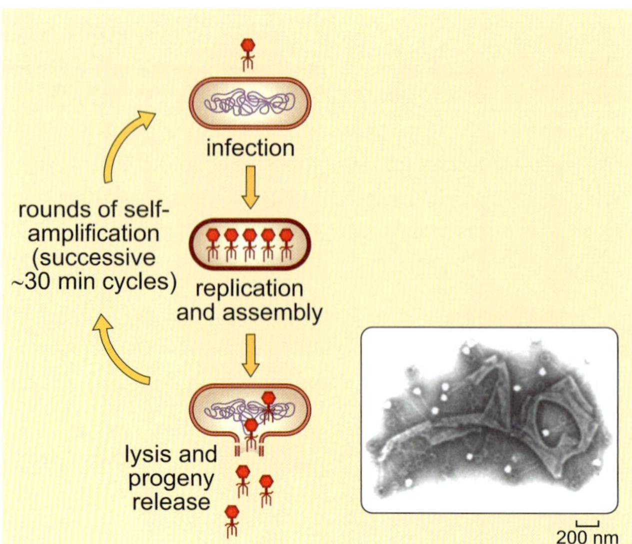

Figure 4.12 Phage therapy exploits the natural lytic cycle, which occurs over about 30 minutes and is divided into three major steps: infection, replication and assembly, and finally the release of new virions (red). Subsequent infection of new host bacteria leads to the process of self-amplification. The electron micrograph depicts phage particles adhering to the debris of a lysed streptococcal cell.

The use of bacteriophages has several potential advantages over other antibacterial therapies:

- Bacteria infected by lytic phages are lysed and therefore killed. This is in marked contrast to many antibiotics which are merely bacteriostatic: they prevent growth, but the continued exposure of viable bacteria to the antibiotic increases the probability of resistance evolving.

- Phage therapy is 'auto-dosing': as each wave of infecting bacteria is lysed, new phages are released to continue the good work. In theory, only a single initial low dose of the original phage preparation would be required, saving significant time and money.

- Phages are non-toxic to the cells of the infected individual.

- The specificity of the phage means that other normal flora (in the gut, for example) is not affected. In contrast, broad-spectrum antibiotics tend to wipe out a great majority of these commensal microbes, potentially allowing opportunistic pathogens to colonise the treated individual.

- Unlike antibiotics, phage resistance is rare.

- Phages can be easily isolated from the environment – anywhere there are high concentrations of bacteria, there will also be a massive variety of bacteriophages.

- Phages can be administered in a variety of ways, from liquids and creams to impregnation of solid surfaces. They can even be combined with antibiotics.

- Phages can clear biofilms much more effectively than antibiotics.

Given this multitude of advantages, it is surprising that phage therapy has not been more widely adopted. There are some difficulties: it is necessary to select phages with certain characteristics. For example, therapeutic phages must be obligately lytic in their life cycle (Section 3.3.2). Viruses, capable of a lysogenic life cycle may be unsafe to use as they may be able to insert their own DNA into the DNA of the host bacterium; this could confer so called 'superinfection immunity' which converts phage-sensitive bacteria into phage-resistant ones.

Selected phages must have low potential to transfer bacterial virulence factors between bacteria. They should also interact to the least possible extent with the patient's immune system and, of course, display good killing potential for the target bacteria. However, possibly the greatest issue facing the wider deployment of phage therapy is lack of familiarity with phages as antibacterial agents in the Western medical establishment. Only a few phage-based therapies have passed regulatory standards for human use, although they are widely used for veterinary, agricultural and food microbiology applications.

When phage therapies were first trialled in the 1940s and 1950s, a lack of understanding of phage biology and various technological limitations resulted in only partial success. This coincided with the start of the antibiotic age, when new and highly successful antibiotics were being discovered and synthesised constantly. With the threat of wide-scale antibiotic resistance some way into the future, there was no pressing need for the pharmaceutical companies to earmark research and development resource into alternative therapies for bacterial disease. However, the rise of multi-drug resistant bacteria is changing this view. In recent years, well-controlled animal trials have shown that phages can limit very serious bacterial infections. The time seems to have come to revisit this interesting approach to controlling bacterial disease. You will investigate the advantages and disadvantages of modern phage therapy in Activity 4.1.

4.2.8 Confusing bacteria: harnessing quorum sensing

For many years it was thought that bacteria acted as individuals, indifferent to the other bacteria around them. It is now known that bacteria can communicate with each other, both with their own species via quorum sensing (Section 2.3.5) and with other species using **quorum sensing crosstalk**. You met an example of both in Section 3.2.3 when you considered the activities of bacteria within biofilms. Bacteria secrete signalling molecules called autoinducers; the more cells there are, the greater the accumulation of the autoinducer. These signalling molecules bind to receptors on the surface of the bacteria, and as the concentration of signalling molecules in the neighbourhood increases, specific genes are switched on or off in all the bacteria. This type of inter-cell signalling allows co-ordination of bacterial activity within the biofilm by regulation of gene expressions systems. Pathogenic bacteria can, for example, delay switching on virulence genes until their cell density reaches a certain level. This prevents the immune system from being alerted and 'picking off' the invading bacteria while their numbers are still small.

As mentioned in Chapter 3, quorum sensing plays a very important role in the virulence of *S. aureus*. This bacterium makes use of a quorum sensing system involving an autoinducing peptide (AIP) encoded by the **a**ccessory **g**ene **r**egulator locus (agr). This complex quorum sensing system is beyond the scope of the module, but suffice it to say that AIP signalling controls over 70 genes, 23 of which code for virulence factors that promote attachment to the host, immune system evasion and toxin production. Activation of these genes effectively converts the bacterium into an invasive and aggressive pathogen.

E. coli also uses quorum sensing to control virulence genes. Recall from Chapter 3, that enteropathogenic *E. coli* attach themselves to the gut wall in order to establish an infection; quorum sensing controls this process. Enteropathogenic *E. coli* carries a large region of chromosomal DNA encoding the virulence factors responsible for lesion formation, including an adhesin called intimin and an intimin receptor (Tir) (Section 3.2.2). The expression of these genes is quorum sensing-dependent and allows the bacteria to establish themselves in micro-colonies on the gut wall.

If alternative autoinducer signalling molecules could be identified, it may be possible to use them to downregulate virulence genes and hence reduce the pathogenicity of these pathogens. In this way, quorum-sensing behaviour could be harnessed as an alternative therapy to combat infectious diseases.

As the threat of antibiotic resistance spreads, it is essential to identify new ways to control and combat infectious diseases. Activity 4.1 will allow you to further explore some of these options.

Activity 4.1 Alternatives to antibiotics: combating multiresistant bacteria

(LOs 4.1 and 4.2) Allow 60 minutes to access the online resources, video and to read the associated notes and answer the questions in this activity

In this activity you will access a number of online resources to further develop your understanding of novel antibacterial therapies, including phage therapy and use of quorum sensing, and learn how new techniques can reduce the risks from the growing threat of antibiotic resistance.

Summary of Section 4.2

- Antibiotics are complex chemicals that block the active sites of specific proteins (and some nucleic acids), including those required for growth.

- Antibiotics are used to control infectious disease in a variety of health and veterinary-related fields.

- Antibiotics are classified according to their chemical structure. Those in the same class have similar or identical core structures but different substituent groups.

- In bacteria, antibiotics may target cell-wall-synthesising enzymes or the machinery for nucleic acid and protein synthesis.

- Bacteria can become resistant to antibiotics by a variety of different mechanisms.

- Resistance genes encode new proteins that underlie each resistance mechanism. These genes arise by mutation of existing genes and are selected for by the widespread use of antibiotics. Horizontal gene transfer can spread these genes.

- Although the pharmaceutical industry is investing hugely in the synthesis of new antibiotics (e.g. Daptomycin), resistance is still a significant threat to any new antibiotic marketed.

- Alternative antibacterial therapies are required to reduce dependence on antibiotics in the future. Phage therapy and the isolation and use of autoinducer signalling molecules to prevent bacterial quorum sensing may provide additional weapons in the fight against bacterial disease.

4.3 Topic 2: Recombinant DNA technology – manufacturing human proteins

In this topic you will focus on the biology involved in engineering bacteria and yeasts to synthesise useful molecules, particularly proteins, using insulin production as an example. This will be followed by a brief exploration of the technology required to grow cells in culture on an industrial scale.

4.3.1 Insulin

Insulin was the first human protein to be manufactured using recombinant gene technology, a major milestone in the application of genetic engineering to solve pharmaceutical problems. Insulin is a regulatory peptide hormone (composed of 51 amino acids), synthesised in the pancreas. Its role is to control the uptake of glucose from the blood (glucose homeostasis). In the absence of insulin, glucose remains in the bloodstream and is unavailable to metabolising cells for respiration. After a meal, the increase in blood glucose levels triggers the secretion of insulin from endocrine cells called β cells in regions of the pancreas called the islets of Langerhans. Insulin travels around the body in the bloodstream and binds to receptors on the surface of cells in target tissues, mainly muscle, adipose tissue and the liver. The binding of insulin allows these cells to take up glucose from the blood and either utilise it for the cell's immediate metabolic needs or convert it to the carbohydrate glycogen for storage. As a result, the level of glucose in the blood falls. Failure to secrete insulin, or an inability of the target cells to bind or respond to insulin, results in the metabolic disorder, diabetes mellitus. In 2011 an estimated 346 million people (5% of the world's population) had this debilitating disease, and the number is increasing inexorably year on year. Predictions suggest that by 2030, 10% of the world's population will be affected.

Diabetes exists in two main forms: type 1 and type 2. The majority of cases are type 2 diabetes (sometimes called late onset diabetes) resulting from the development of insulin resistance where target cells are unable to respond to insulin. Both genetic and lifestyle factors, including diet and obesity, and also age have been linked to the likelihood of developing the type 2 disease. Type 1 diabetes, also known as insulin-dependent diabetes, accounts for 5–15% of cases; the hallmark of this form of the disease is an inability to secrete insulin as a result of loss of β cells from the pancreas. Such individuals experience chronic hyperglycaemia (elevated blood sugar). This causes symptoms such as increased hunger and thirst, but can lead to so-called diabetic emergencies – including the life-threatening diabetic ketoacidosis, a condition which occurs when the body cells switch from metabolising glucose to fatty acids leading to a significant decrease in the pH of the blood. The long-term effects of this disease include multiple organ damage, blindness, kidney failure, heart disease, stroke, neuropathy, and amputations. The mainstay therapy for type 1 diabetes is insulin treatment administered several times daily.

4.3.2 Therapeutic insulin

Therapeutic insulin has been available commercially since 1921 when Canadian scientists Frederick Banting and Charles Best first isolated insulin from the pancreas of dogs.

By 1923, the pharmaceutical firm Eli Lilly were mass-producing insulin but their method of harvesting insulin from the pancreases of dead animals (bovine and porcine insulin) was not sustainable in the long term; the demand for insulin therapy was simply too high.

With the advent of gene cloning, the potential of using cultured cells, particularly bacteria, to manufacture human proteins became apparent. Recall from Book 1, Box 5.2, the methods by which genes can be transferred between species. The principle of gene cloning relies on taking a piece of DNA derived from an organism, inserting the DNA into a bacterial cell and then providing the bacterial cells with the necessary conditions to grow and multiply. If the cloned DNA encodes a protein, the bacterial cells will be able to synthesise the protein, provided that the DNA sequence is in the appropriate form. As a result of this technology, human therapeutic proteins can be manufactured on a much larger scale than ever before. Insulin production was the first human protein to be industrially produced using this technique.

The first step was to identify the complete DNA sequence of the human insulin gene. Once identified however, there are some difficulties in simply inserting the human gene into bacterial cells in the hope that the cells will make active insulin protein. Prokaryotic and eukaryotic genes differ at a fundamental level; eukaryotic genes contain introns, non-coding sequences that are spliced out of the primary transcription product to produce a mature mRNA (Book 1, Chapter 6). Prokaryotic genes in contrast consist of a continuous coding region that, when transcribed, produces a mature mRNA that has no requirement for post-transcriptional modification. While bacteria can perfectly well transcribe eukaryotic DNA, they cannot therefore process the primary transcript to splice out introns, so a suitable DNA sequence without introns must be provided.

A second issue concerns the relationship between the primary product of translation of the insulin mRNA sequence and the mature active insulin molecule. Figure 4.13 shows the structure of the active insulin molecule, which consist of two polypeptide chains, the A chain (21 amino acids in length) and the B chain (30 amino acids in length). These two chains are joined together by two disulfide bonds; recall these are covalent links that form between the –SH side chains of two cysteine residues (Book 2, Section 1.5.1). An additional disulfide bond between two cysteine residues within the A chain also contributes to the final conformation of the insulin molecule.

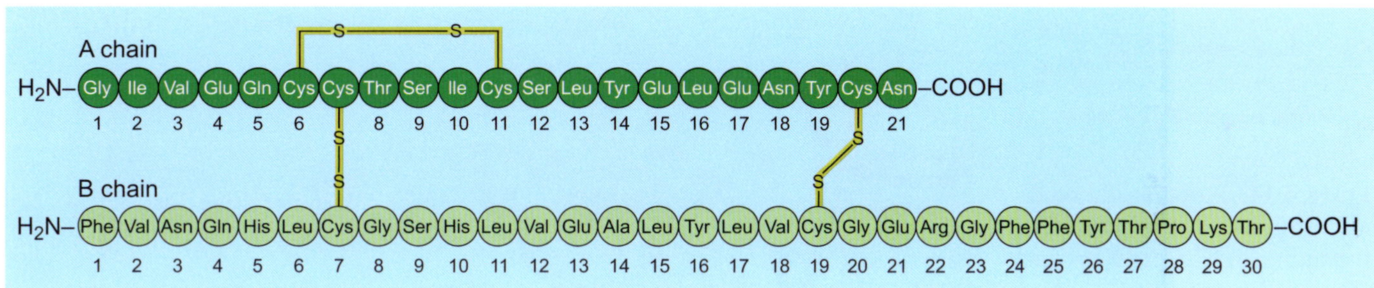

Figure 4.13 Representation of the amino acid sequence of the active human insulin molecule showing, the A and B chains linked together by disulfide bonds between cysteine residues. Note that a disulfide bond also forms between two cysteine residues within the A chain.

The structure of active insulin is significantly different from the initial translation product. In Book 2, Section 1.6.2 you were introduced to examples of proteins that were converted into their active form by post-translational modification. Recall that insulin was one of these proteins; requiring post-translational proteolytic cleavage to convert the inactive precursor protein to active insulin. The initial translation product of the insulin gene is preproinsulin, a single polypeptide consisting of 108 amino acids. As well as the sequences of the A and B chains, preproinsulin includes two further sequences: the C chain or connecting peptide (33 amino acids long), which links the A and B chains, and a 24 amino acid long signal sequence or 'tail' at the N-terminal end of the B chain. Figure 4.14 summarises the post-translational cleavages preproinsulin must undergo to produce active insulin.

These additional amino acid sequences need to be removed in order to produce fully functioning insulin; however, they play an important role in allowing the active hormone to achieve the correct final conformation and to be secreted from the cell. The position of C peptide ensures that the A and B chains are placed in the correct configuration to bring the cysteine residues in close proximity to form disulfide bonds and the N-terminal signal sequence helps direct the molecule to the rough endoplasmic reticulum (RER) of the β cell.

■   Why does the proinsulin need to be directed to the rough endoplasmic reticulum (RER)?

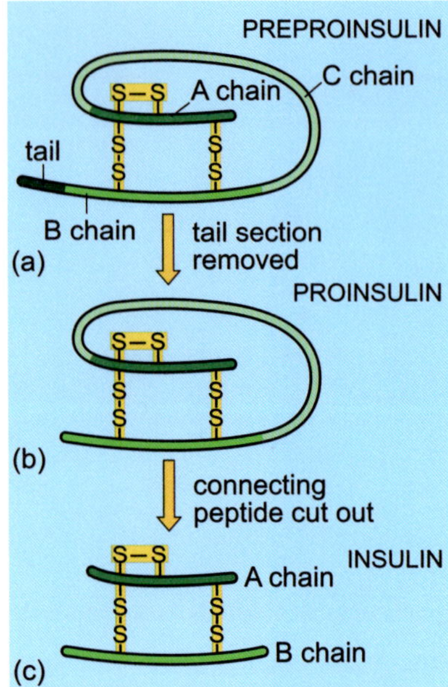

Figure 4.14 Post-translational modification in the production of insulin. (a) The gene product is the insulin precursor, preproinsulin. (b) Removal of a short 'tail' section produces proinsulin. (c) Excision of the connecting peptide (i.e. amino acid sequence), also called the C chain or peptide, leaves the mature insulin molecule, i.e. the A chain and the B chain joined together via two disulfide bonds.

□ The rough endoplasmic reticulum (RER) is the entry point into the secretory pathway for newly synthesised proteins destined for the cell surface or to be secreted. Because insulin needs to be secreted in order to circulate in the blood, it is essential that after synthesis, it finds its way into the RER (Book 1, Section 3.4.8).

Figure 4.15 shows the sequence of events from transcription of the insulin gene to the final secretion of mature insulin from the endocrine cells of the pancreas.

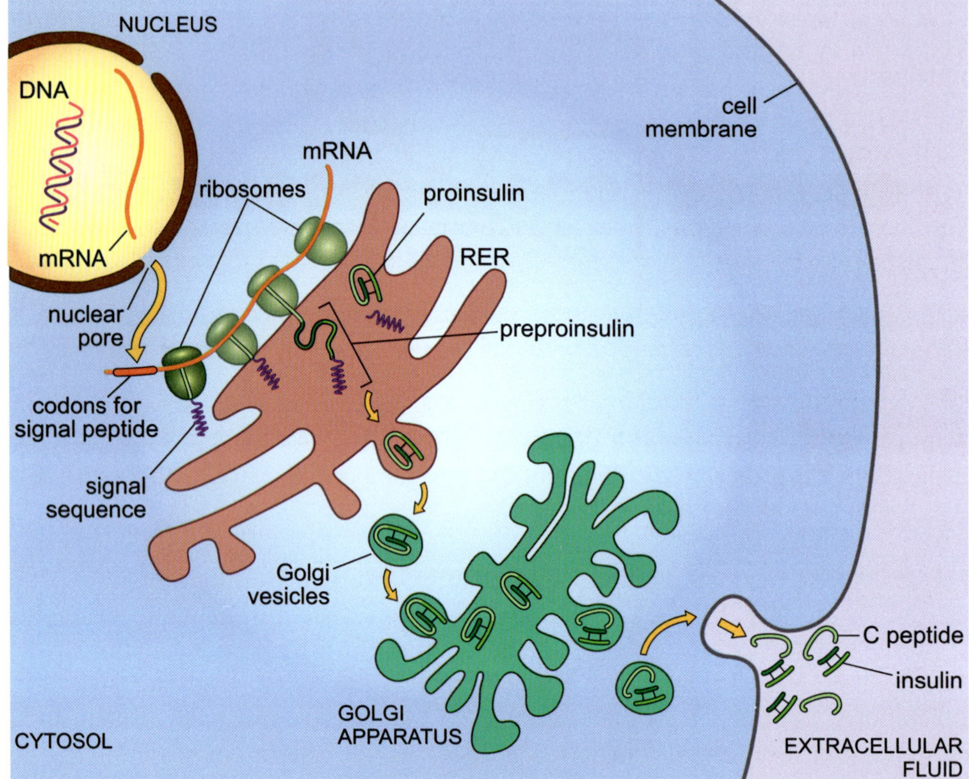

Figure 4.15 The sequence of events from transcription of the insulin coding sequences to the final secretion of mature insulin in the β cells of the pancreas.

As soon as the mRNA starts to be translated by cytoplasmic ribosomes, the N-terminal signal sequence of preproinsulin is bound by a signal recognition particle and transported to the rough endoplasmic reticulum where the preproinsulin is translocated into the RER lumen (Book 1, Section 3.4.5). A protease in the RER lumen then cleaves off the signal sequence from preproinsulin to leave proinsulin. Still in the lumen of the RER, the proinsulin folds into its final conformation, stabilised by the three disulfide bonds that form under the control of enzymes such as protein disulfide isomerase. Protein folding in the RER is also facilitated by the action of molecular chaperones (Book 2, Section 1.5.3). The RER of a pancreatic β cell produces approximately 1 million molecules of insulin a minute, so it is important that synthesis of preproinsulin by the ribosomes and folding of the protein in the RER keep up with each other. Despite the presence of chaperone mechanisms,

a significant amount of proinsulin is misfolded within the RER of β cells –
this is typical of protein production in all cell types and such proteins must be
dealt with by a 'quality control system'. If the capacity of this system is
exceeded and misfolded proteins begin to build up in the RER, then a stress
response is switched on.

■ Why are incorrectly folded proteins useless to the cell?

☐ The folding of the protein into its correct secondary and tertiary structure
determines its function (Book 2, Section 1.5).

Misfolded but soluble proteins are translocated out of the RER for ubiquitin-
dependent degradation by the proteasome (Book 2, Section 1.5.4). Insoluble
misfolded proteins remain in the RER and may be eliminated by autophagy:
destruction of internal cell components, in this case, parts of the RER, by the
action of lysosomes (Book 1, Section 3.4.9). It is possible that the origin of
type 1 diabetes itself lies in the malfunction of these regulatory mechanisms.
The inability to clear misfolded and accumulated protein aggregates may lead
to apoptosis (Section 1.3 of this book) and therefore loss of β cells.

■ Where else in the module have you met with disease linked to the
accumulation of misfolded protein?

☐ Accumulation of misfolded proteins is a characteristic of cystic fibrosis,
Alzheimer's disease, Huntington's disease, and Parkinson's disease
(Book 2, Section 1.5.5 and Chapter 3 of this book).

Correctly folded proinsulin is transported in vesicles pinched off the RER to
the Golgi apparatus (Book 1, Section 3.4.7), where another protease cleaves
the C chain from proinsulin to produce the active form of insulin which is
packaged into secretory granules. Insulin and C peptide are secreted from the
cell when secretory granules fuse with the cell membrane in the process of
exocytosis (Book 2, Section 2.9.1).

In summary, in addition to being capable of transcribing and translating the
insulin gene, pancreatic β cells are able to process and package insulin
appropriately and transport it outside the cell. Ideally, cheap and efficient mass
production of insulin, in other types of cells requires a cell system which
mimics these events as far as possible. Bacteria have limitations in this
respect. As well as differences in transcription, the processes of post-
translational modification seen in eukaryotes are absent in bacteria.
Consequently, other eukaryotic organisms, including yeasts, are now
commonly used for insulin production. However, because bacteria were the
first genetically engineered insulin-producing cells, you will study how they
were developed before considering solutions involving the use of eukaryotic
cells.

4.3.3 Manufacturing therapeutic insulin: gene cloning in bacteria

In this section you will consider the first step in the manufacture of human
insulin: 'cloning' the human gene in bacteria. Recall that the bacteria would

have to be provided with an intron-free DNA sequence encoding preproinsulin.

■ How might researchers obtain an intron-free preproinsulin sequence?

□ They could isolate preproinsulin mRNA from human cells (which the cells have processed to remove introns) and use it as a template to make a DNA copy (Book 1, Box 6.2).

■ Name an enzyme capable of synthesising DNA from an RNA template.

□ Reverse transcriptase (obtained from a retrovirus).

In the 1970s, two rival research groups (Rutter and Howard at the University of California and Gilbert at Harvard) took just this approach and isolated purified mRNA from the β cells of hundreds of rat pancreases in one case, and from cultured rat tumour cells in the other. Isolating the correct mRNA required separation of all the mRNAs produced in the β cells using the technique of gel electrophoresis (Book 1, Box 6.2).

■ How can the correct mRNA be identified?

□ The correct mRNA was identified in the gel by its size, since the researchers knew they were looking for an mRNA of a specific length.

Once the mRNA template had been isolated, reverse transcriptase was used to synthesise a single-stranded DNA (ssDNA) copy complementary to the mRNA (referred to as complementary DNA or cDNA). DNA polymerase was then used to convert the ssDNA to a double-stranded DNA (dsDNA).

At the same time, a third research group at Stanford University (Walter Boyer and Stanley Cohen) approached the problem of producing intron-free insulin DNA in a different way. They used the genetic code to work out the DNA sequence based on the known amino acid sequences of the A and B chains of mature insulin. Two so-called artificial genes, one for each chain, were then created by chemical synthesis of ssDNA molecules from nucleotides *in vitro*, and then once again using DNA polymerase to convert the ssDNA to dsDNA.

Superficially, this seems a much more straightforward approach than working back from mRNA to cDNA as the other research groups did. Making separate A and B chains circumvents the need for the protein processing that would be required to convert preproinsulin to active insulin (Figure 4.14) and it was far easier to work with artificial DNA, since the restrictions placed on groups working with human DNA at the time were extremely onerous. Actually, it was fortunate that the A and B chains in insulin were such short amino acid sequences, since accurate chemical synthesis of long sequences of DNA was at that time actually very challenging.

Hence both the cDNA and artificial DNA methods produce intron-free DNA, but how can bacterial cells be induced to take up a piece of foreign DNA, to make copies that will be passed on when the cells divide, and to synthesise the encoded protein? In Book 1, Box 5.2 you were introduced to the concept of gene cloning and the importance to this process of two naturally occurring features of bacterial cells: plasmids and restriction enzymes. Without the existence of either of these two features, gene cloning would not have been possible. Plasmids are circular dsDNA molecules that can replicate independently inside cells, and can make up as much as 25% of the genetic material of some bacteria (Figure 4.16).

- What form does the rest of the genome of bacteria take?

☐ The bacterial genome usually consists of a larger closed circle of dsDNA: a single chromosome.

Plasmids are the cloning vectors most commonly used to introduce 'foreign' DNA into bacteria and other types of cell. However, in order to do this, there needs to be a mechanism of opening up plasmids, inserting the required DNA and then resealing them.

- Briefly outline the process used to insert DNA into a plasmid.

☐ Restriction enzymes are used to cleave plasmid DNA at specific sequences, leaving cohesive ends; the DNA fragment of interest, derived by cleavage with the same enzyme, anneals to the cleaved plasmid and the DNA ends are joined using DNA ligase, producing a recombinant plasmid (Book 1, Box 5.2).

4.3.4 Making the recombinant insulin plasmids

The plasmid used for the very first insulin gene cloning procedure was called pBR322. Book 1, Figure 5.18 shows the structure of typical plasmid cloning vector, illustrating the three essential features: an origin of replication, one or more selectable markers (in the case of pBR322, genes conferring tetracycline resistance and ampicillin resistance) and one or more cloning sites (for pBR322 there are four sites where different restriction enzymes can cut the plasmid to allow insertion of a DNA fragment).

- Why does the plasmid require an origin of replication?

☐ The origin of replication is the site on the plasmid where replication of the plasmid DNA is initiated. Plasmids need to be replicated in the cell, otherwise they will be lost as the cell divides.

Generally, any plasmid will replicate itself at least once in every cell cycle, thereby maintaining the copy number of that plasmid in subsequent generations of cells, and retaining the characteristics the plasmid confers on the cell. For gene technology it is vital that recombinant plasmids are retained within the cell line long-term, or the yield of useful product expressed from the plasmid will fall over time.

After insertion into cells, it is necessary to be able to distinguish those cells that contain the plasmid vector from those that do not. There is little point culturing cells that do not contain the gene required to synthesise the desired

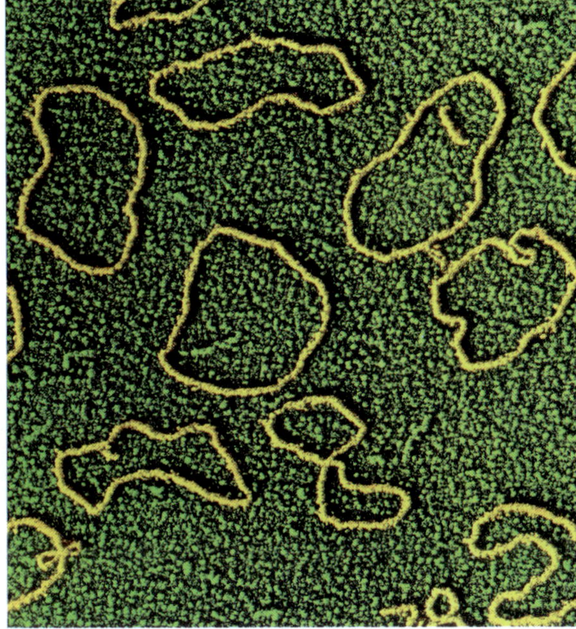

Figure 4.16 False-colour transmission electron micrograph (TEM) of plasmids of bacterial DNA from the bacterium *E. coli*. This plasmid, designated pBR322, is frequently used in genetic engineering work. DNA is inserted into it and the composite plasmid is then introduced into an *E. coli* bacterium where it propagates and expresses the genetic information carried by the transplanted DNA.

product. The presence of selectable markers on plasmids used for genetic engineering is therefore essential. The selectable marker must change the phenotype of the host cell in a way that allows the presence of the vector to be detected. The most common method uses a gene that confers antibiotic resistance, for example, the gene encoding resistance to tetracycline, *tetR*. Only those cells carrying the antibiotic resistance gene will actually grow in media containing that antibiotic.

■ Can you foresee any problems with using antibiotic resistance genes as markers?

□ If the recombinant bacterium is released into the environment, it is possible that antibiotic resistance might be transferred to a pathogenic bacterium (Section 3.2.3).

Once a recombinant plasmid has been constructed, the plasmid DNA is incubated with suitable 'competent' *E. coli* bacteria that take up the naked plasmid DNA. Cells that have taken up plasmid DNA are said to be transformed (Book 1, Section 5.7.5 and Box 5.2). The transformed cells are spread on agar plates containing antibiotic in order to selectively grow only those bacteria carrying the recombinant plasmid.

4.3.5 Insulin expression in bacteria

In order for a cell to synthesise a protein, its gene has to be transcribed and translated and you will remember from Book 1, Chapter 6 that gene expression is subject to complex control mechanisms.

■ What steps initiate and terminate transcription of prokaryotic DNA into mRNA?

□ RNA polymerase binds to the DNA molecule at a region called the promoter, a highly conserved DNA sequence a fixed number of bases upstream of the transcription start site. Once the RNA polymerase binds to the promoter, transcription proceeds from the start site, along the DNA molecule until the polymerase encounters a transcription terminator sequence (Book 1, Section 6.3).

It is therefore essential that the insulin DNA-bearing plasmid contains a suitable promoter sequence. The race to be the first to express recombinant human insulin was actually won by the Boyer and Cohen group who used the *lac* operon promoter (Book 1, Section 6.3.3) to drive the expression of their artificial DNA molecules coding for the insulin A and B chains. A section of *lac* operon DNA, including the *lac* promoter and the *lacZ* gene which encodes the β-galactosidase protein, was inserted into the plasmid, immediately upstream of the DNA sequence encoding either the insulin A chain or the B chain, such that the recombinant plasmid encoded a fusion protein with β-galactosidase at the N-terminus and the insulin A or B chain at the C-terminus. Figure 4.17 shows the process used by Boyer and Cohen. The complete human preproinsulin gene (derived from mRNA) was later expressed in bacteria in a similar way.

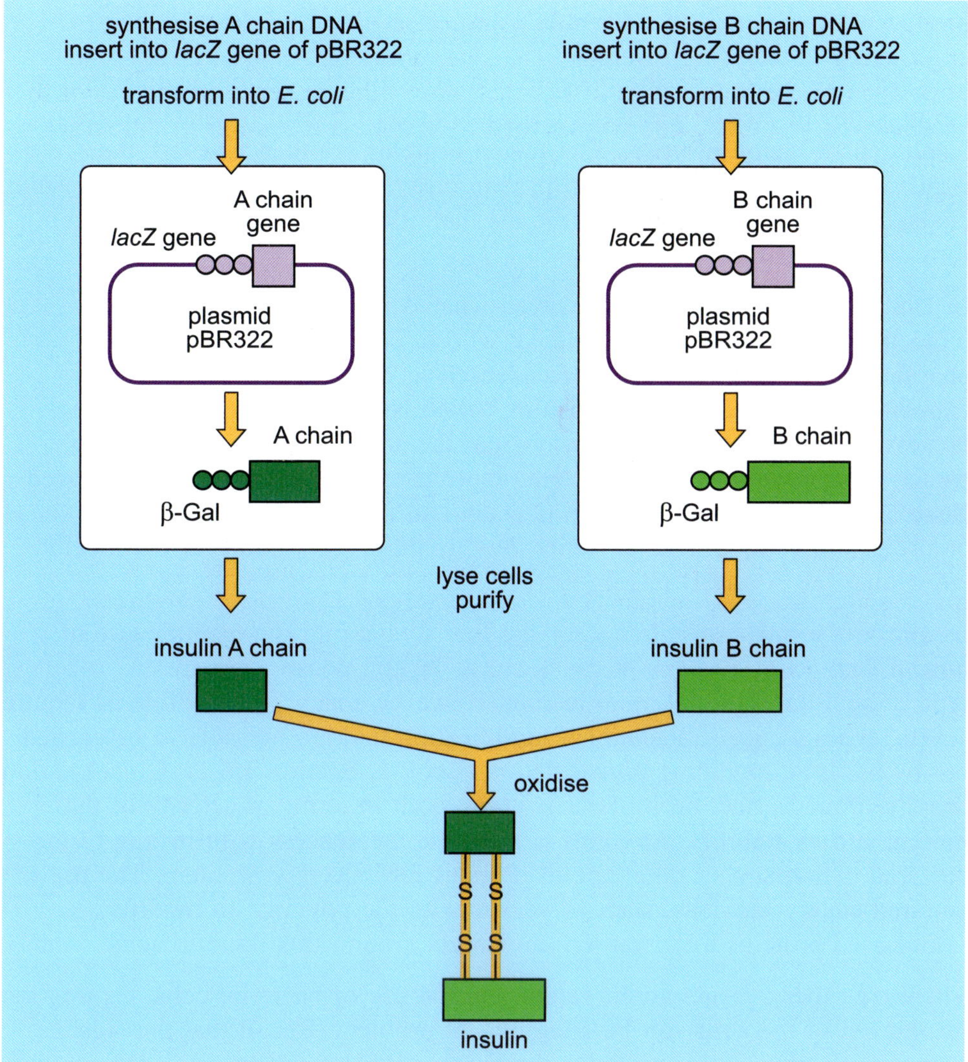

Figure 4.17   The Boyer and Cohen method for human insulin production. The artificial genes for the insulin A chain and the insulin B chain are separately inserted into the *lacZ* gene (encoding β-galactosidase) of two pBR322 plasmids. The plasmids are transformed separately into populations of *E. coli*. The bacteria synthesise fusion proteins consisting of β-galactosidase–chain A and β-galactosidase–chain B. The A and B peptides are cleaved from the B-galactosidase portion and then purified and chemically combined to give active insulin.

■   Under what circumstances will the *lac* operon promote the transcription of the hybrid β-galactosidase–insulin chain gene?

☐   The presence of lactose in the growth medium induces the *lac* operon to promote the transcription of the *lac* operon genes, or in the case of this recombinant plasmid, the hybrid β-galactosidase–insulin chain gene.

4.3.6 The final product: manufacturing active insulin

E. coli transformed with the plasmid and grown under antibiotic selection in the presence of lactose produces large amounts of hybrid β-galactosidase–insulin chain fusion proteins. The fusion proteins are then isolated, the β-galactosidase part of the fusion protein cleaved off, and the A and B insulin chains mixed together under conditions that allowed the disulfide bonds to form between them.

In 1982, Boyer and Cohen established what is now one of the largest biotechnology companies, Genentech, who in conjunction with the pharmaceutical firm Eli Lilly were licensed to manufacture the bacterially-expressed recombinant human insulin, called Humulin. By the mid-1980s however, production switched away from the artificial DNA technique to bacterial expression of the complete preproinsulin gene. Ultimately it proved easier and cheaper to remove the C chain and the signal sequence from preproinsulin using proteases rather than going to the trouble of chemically combining the A and B insulin chains.

Insulin was a fortuitous choice for the first forays into the manufacture of human therapeutic proteins because active insulin doesn't require any other type of post-translational modification. However, many human proteins require several types of post-translational modification that are not able to be carried out by bacteria, including glycosylation (Book 2, Section 1.6.1). Glycosylation has a profound effect on protein function, both in terms of activity of the protein and its stability. Attempts to translate the success with insulin to the bacterial expression of other useful proteins ran into this problem. The use of cultured eukaryotic cells, such as yeasts, with the potential to modify synthesised proteins after translation can get around this issue to some extent.

There are other considerations relating to the use of bacterial cells. Getting the insulin out of the cells can be a significant problem. Recall that the signal sequence on preproinsulin allows the pancreatic islet cell to direct the protein to the cell's RER and Golgi apparatus, enabling it to be secreted from the cell. However bacterial cells do not recognise the human signal sequence, and consequently the protein builds up inside the cell cytoplasm as insoluble inclusions. The cells must be disrupted by high-pressure homogenisation and the cytoplasmic inclusions must then be separated from the cell debris and dissolved in a solvent. This step, and the additional downstream processing (Section 4.3.8) of the protein product to obtain its active form, add considerable costs to the industrial process.

4.3.7 Using yeast for the manufacture of proteins

Yeasts are highly cost-effective alternatives to bacterial cells and share many of the advantages of bacteria: they can be easily genetically manipulated and grow rapidly on simple media; however, an advantage yeasts have over bacteria is that they are capable of the protein processing (including glycosylation) and the protein targeting typical of all eukaryotic cells.

Eukaryotic expression vectors have similar characteristics to bacterial expression vectors. They require an origin of replication, suitable cloning sites for insertion of the DNA of interest, a selectable marker of some description (which usually confers the ability to grow in a particular selective medium), a suitable eukaryotic promoter, and additionally a sequence that will polyadenylate the mRNA (Book 1, Section 6.5.1).

One of the great benefits of the yeast system for manufacturing human proteins is the yeast cell's ability to secrete the protein into the growth medium. However, yeast will only do this if the protein has a signal sequence capable of directing the protein to the RER – just as in the secretion of insulin by pancreatic β cells. To this end, scientists have made use of an amino acid sequence that yeasts naturally secrete, the alpha factor mating pheromone. This is a short peptide signalling molecule that advertises the presence of an alpha type yeast cell to neighbouring cells of a different mating type (Book 2, Section 4.1). In order to facilitate the secretion of a recombinant protein synthesised in yeast, the gene encoding the precursor of the mating factor is cloned in front of the cDNA for the desired protein. This is known as a 'leader' peptide and its presence relocates translation of the fusion protein to the RER, where it then passes to the Golgi apparatus and is ultimately secreted. The leader peptide is removed naturally by a yeast endoprotease before secretion. Insulin is frequently manufactured using this technique by expression in *Saccharomyces cerevisiae* of a smaller insulin precursor (IP), comprising just the first 29 residues of the insulin B chain and the 21 residues of the insulin A chain connected by the peptide sequence Ala–Ala–Lys, as a single fusion protein. The fusion protein is then harvested from the culture medium. Treatment with the protease trypsin cleaves the two chains apart and further processing converts the product into active human insulin.

The yeast *S. cerevisiae* was first used on an industrial scale for the production of a recombinant protein in the 1990s. Currently, about 40% of all therapeutic proteins are synthesised in either *E. coli* or *S. cerevisiae,* and some non-glycosylated proteins (like insulin) are successfully expressed in both systems; for example, recombinant insulin for clinical use is produced in both *E. coli* (Humulin, produced by Eli Lilly in the USA) and yeast (by NovoLog in Denmark). In general, production in yeast cells is usually favoured when the target protein requires post-translational modification prior to secretion in a soluble form.

The use of *S. cerevisiae* for protein expression does have some drawbacks however, including loss of plasmids, low yields and also hyperglycosylation – the addition of too many carbohydrate chains to some secreted proteins, which can affect the activity of the protein or alter its immunogenicity. The yeast *Pichia pastoris* produces higher yields and less hyperglycosylation and is now commonly used as an alternative to *S. cerevisiae.* Expression in other types of eukaryotic cells, including cultured mammalian cells, is sometimes used for manufacture of proteins that are more difficult to obtain in an active form. Regardless of the choice of organism, in order to manufacture the protein of interest, it is necessary to optimise the culture of cells on a large scale, harvest the synthesised protein and purify it. This is the subject of the next section.

4.3.8 Process technology

Process technology is the term used to describe the technology required for commercial production of cell products in large culture vessels called bioreactors (sometimes confusingly referred to as 'fermenters', even though the process taking place is not anaerobic fermentation). When setting up cell culture on a commercial scale, it is vital to first establish the optimum growth conditions for the recombinant cells.

- ■   What factors may be significant to the growth rate of cells grown in culture?

- □   Oxygen availability and nutrient supply, pH and temperature are all factors that affect cell growth.

The optimum conditions for growth of the cells are first established by trial and error using small-scale cultures grown in laboratory flasks. However, maintaining these conditions at the much larger scale required for commercial production poses quite a challenge for biochemical engineers. In addition, keeping unwanted bacteria or fungi out of these large volumes of culture is much more difficult than it is on a small scale.

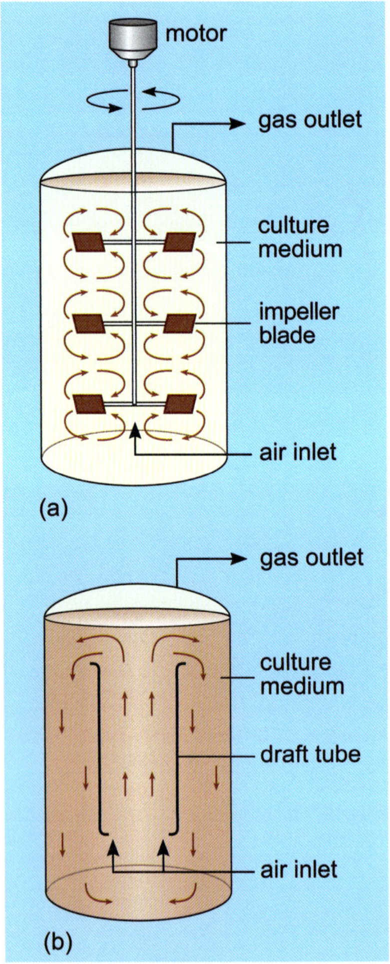

When scaling up production, a major challenge is to provide the growing cells with sufficient oxygen. In the laboratory, simply stirring or shaking the culture vessels can achieve adequate aeration. On an industrial scale, continuous mechanical stirring is required and air or even pure oxygen may be bubbled through the much larger volume of culture medium, often at high pressure. Bubbling air through the culture has the added benefit of mixing the culture medium: this disperses nutrients, heat and waste products. Sometimes, the growing cells release so much heat energy that additional cooling of the culture is required. Large cooling coils containing circulating cold water are placed within the culture vessel to ensure the medium is maintained at the optimum temperature for cell growth. Figure 4.18 shows a stirred tank bioreactor and an air injection bioreactor.

Essential in all types of cell culture is the need to maintain a pure culture; if bacteria or yeast from the atmosphere, or the growth media or the inside of the culture vessel contaminate the bioreactor, they will potentially outcompete the cells of interest and the yield of useful product will be reduced. Genetically engineered cells frequently have lower growth rates than other cell types; the large number of plasmid copies they carry imposes a significant genetic burden and reduces the rate of DNA replication. Consequently, contaminating cells from the environment can quickly swamp the culture vessel. This means that all liquid media, vessels, and entry and exit points from the culture vessel have to be kept sterile. This requirement significantly raises the costs of the industrial-scale process.

motor

gas outlet

culture medium

impeller blade

air inlet

(a)

gas outlet

culture medium

draft tube

air inlet

(b)

Figure 4.18 (a) A stirred tank bioreactor and (b) one type of air injection bioreactor. The brown arrows denote the pattern of liquid circulation.

Batch or continuous culture?

Some large-scale biotechnological syntheses are performed in **batch culture** mode, while others are run as **continuous culture** processes. In addition to these two extremes is the **fed-batch culture** process. Figure 4.19 shows the relationship between the concentration of growth medium (substrate) and cell concentration with time for each of these three modes of operation. The characteristics of each mode are summarised here.

Batch culture: Once the culture is added, nothing is added to or removed from the bioreactor during the period of culture. As shown in Figure 4.19a, there are the three phases of growth: lag, exponential and stationary (described in Section 1.2.1; note that the death phase is not shown here). The concentrations of nutrients decline over time and there are corresponding increases in the levels of waste products. Not only does the cell density (concentration) change but so do cell metabolism and internal composition. Many bacteria synthesise proteases in the stationary phase; so if the required product is a protein, the culture must be harvested before this stage is reached, otherwise yields may be reduced. In contrast, secondary metabolites (products of cellular biosynthesis) such as antibiotics are usually synthesised in the stationary phase, so it is during this phase that the culture is harvested for these products.

Fed-batch culture: Here substrate concentration is kept constant by adding more substrate, a little at a time, at intervals throughout the culture period. The exponential and stationary phases last longer, so that both the cell density and concentrations of bacterial products remain consistently higher than in the batch culture mode (see Figure 4.19b). However, fed-batch processes need more monitoring and control than simple batch systems and so are more costly to run.

Continuous culture: In continuous culture, the composition of the growth medium and the cell density are kept constant (Figure 4.19c) by continuously tapping off the culture and adding fresh growth medium at the same rate.

- ■ What characteristic of the bacteria determines the rate of medium addition and culture removal required to maintain a constant cell density during continuous culture?

- □ Their growth rate – the shorter the cell cycle, the greater the throughput of medium required to keep cell density constant.

There are several advantages of continuous culture over either of the batch modes:

- smaller bioreactors are needed (and correspondingly smaller-scale equipment for downstream processing – see below)
- there is less 'down-time', i.e. time when the bioreactor is not operating; for batch processes, the bioreactor is out of action for cleaning and sterilisation between each batch
- product yields are more consistent.

The constant environment of a continuous culture keeps the physiological state of the cells more uniform throughout the process, making product yields correspondingly uniform. In batch operations, on the other hand, just a small difference in harvest time (i.e. a different point on the cell growth curve) can mean significant differences between batches.

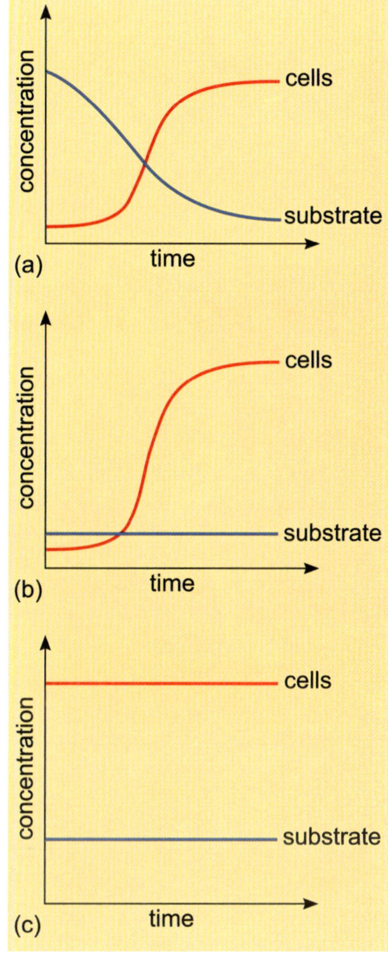

Figure 4.19 Characteristics of (a) batch culture, (b) fed-batch culture, and (c) continuous culture modes. In (b) the transient increases in substrate concentration that occur when substrate is added are too small to be shown at this scale.

However, there are two important disadvantages of maintaining cultures for very long periods (typically 500–1000 hours):

- keeping contaminating bacteria out

- loss of recombinant plasmids; as mentioned earlier, cells without plasmids grow and divide faster than recombinants, so product yield can fall off very quickly over long periods of continuous culture.

Downstream processing

Downstream processing is a collective term for all the operations involved in isolating the product of bacterial metabolism. The first step is to separate the cells from the culture medium, which is usually carried out by one of two methods:

- *centrifugation* – precipitation of the cells by centrifugation leaves the medium (supernatant) almost free of cells; this method is the one most commonly used

- *membrane filtration* – the culture is filtered through a membrane that is porous to the medium but retains the cells.

The next step depends on whether the product is in the culture medium (e.g. insulin production by yeast) or retained within the cells (e.g. Humulin production in bacteria). It is relatively easy to concentrate and purify a protein or a small-molecule product from the medium, but to obtain an intracellular product the cells must first be broken open to release their contents. Quite vigorous methods generally have to be used to rupture the tough cell walls of bacteria, though care must be taken not to denature the product. The choice of method depends on the cell wall composition; there are substantial differences between the cell walls of Gram-positive bacteria, Gram-negative bacteria and yeasts. The composition and therefore the resistance to disruption of a particular bacterial cell wall also vary with such factors as its growth rate and the stage of the cell cycle. Fortuitously, bacteria that are expressing a cloned gene tend to have weaker cell walls and so are ruptured by relatively gentle methods, for example, by transferring the cells to a medium with a lower solute (salt) concentration than the cell contents (Book 2, Section 2.7.1). Water enters the cells by osmosis and the internal pressure causes them to burst. After cell disruption, the insoluble cell debris is removed by further centrifugation or filtration and the product is purified from the solution remaining.

Now that you are familiar with the processes of gene cloning in the context of insulin production, you will learn more about the exploitation of these technologies to mass-produce human proteins, in Activity 4.2.

Activity 4.2 Genetic engineering of therapeutic proteins

(LOs 4.3 and 4.4) Allow 30 minutes to access the online resources, to read the associated notes and answer the questions in this activity

In this activity you will access a number of online resources to further develop your understanding of the role of gene technology in the industrial-scale production of human therapeutic proteins.

Summary of Section 4.3

- Individuals with type 1 diabetes are deficient in insulin, the hormone produced by β cells in the islets of Langerhans in the pancreas. The ability to manufacture this hormone artificially was a landmark event in the use of gene technology to mass-produce useful products.

- In mammals, insulin is synthesised as preproinsulin and then cleaved during post-translational modification to form two chains of unequal length stabilised by disulfide bonds.

- In the β cells of the pancreas, the presence of a signal sequence on the insulin molecule ensures that it is transported to the Golgi apparatus for secretion outside the cell.

- One method used to transfer the insulin gene to bacteria used double-stranded cDNA, obtained from mRNA by reverse transcription. An alternative method manufactured artificial genes for the A and B insulin polypeptide chains.

- The artificial DNA method was first used (by Genentech) to manufacture human insulin (Humulin) on a commercial scale.

- Yeasts are useful alternatives to bacterial cells for the production of recombinant proteins, with the advantage that, like all eukaryotic cells, they are capable of protein post-translational modification (including glycosylation) and protein secretion.

- For any commercial biotechnological process, the cultured cells must be maintained at the optimum temperature, pH, oxygen availability and nutrient levels. The method of mixing, mode of aeration and cooling system must be appropriate to the cultured cell.

- There are three types of large-scale culture systems: batch, fed-batch and continuous.

- Downstream processing is easier if the product is secreted by the cells rather than retained in the cells. For secreted products, the cells must be separated from the medium containing the product and the product purified. For retained products (e.g. Humulin in bacteria) the cells must be disrupted before the product is purified.

4.4 Topic 3: An introduction to stem cell therapy

Gene cloning and biotechnology clearly provide one solution to the problem of insulin-dependent type 1 diabetes by providing safe, high-quality human insulin in large quantities. Since Humulin was first patented, several new analogues of human insulin have been manufactured with small changes to the primary structure of the protein conferring some new characteristics. Patients can be supplied with a mixture of different insulins, some of which are fast-acting and short-lived, while others are long-lasting. Despite these advances in insulin therapy, some scientists have returned to the root cause of the disease to look for a more permanent solution.

If an individual has non-functioning islets cells, why not replace the cells and fix the problem at source? Transplantation of the whole pancreas provides a highly successful outcome for many patients; however, the availability of organs for transplantation rules this out as a mainstay therapy. Transplanting islet cells themselves has also been tried, but just as with any transplantation, even if donor individuals are chosen who are the closest match possible immunologically, the risk of rejection is very high. It is necessary to give immunosuppressant drugs to prevent rejection by the body of these 'foreign' cells. These drugs are essential, but severely limit the useful life of the transplanted cells.

Instead of taking donor cells from other individuals the ideal solution would be to 'laboratory grow' the patients' own cells. From your study of Chapter 1 of this book, you will be aware that once the cells of multicellular organisms have differentiated into specific cell types they are frequently incapable of further cell division. However, specialised non-differentiated cells exist which, via cell division, may allow repair and regeneration of damaged tissue.

■   What type of cell is required to generate new cells capable of differentiation into a specific cell type in animals?

□   Stem cells.

If stem cells could be induced to produce the differentiated insulin-producing β cells that are lacking in type 1 diabetics, then it is possible that this could be a long-term therapy for this debilitating disease.

Diabetes is, however, only one of many disease states which could benefit from stem cell therapy. In this topic you will focus on the manipulation of various types of stem cells in order to provide therapies for a variety of diseases, including the example of a neurodegenerative condition you encountered in Chapter 3 of this book: Parkinson's disease.

4.4.1 A first look at stem cells

Most cells in multicellular organisms are fully differentiated; this means they have a specific morphology and function that cannot change during the lifetime of the cell. However, there are small populations of cells that are undifferentiated. These stem cells are the precursors to all cells in the body, which support crucial renewal and repair processes. In Chapter 1 you considered cell differentiation, the process by which stem cells, develop into specific cell types. There, stem cells were described as either totipotent, pluripotent or multipotent. These terms reflect the relative potential of different types of stem cells to replicate and replace different types of tissue within the body.

- ■ Which type of cells are described as totipotent?

- ☐ The zygote itself and the very first cells derived from it are totipotent. These cells can divide to produce a complete organism.

Human embryonic stem cells (hESCs), the 30–34 cells found in the inner cell mass of the 5-day-old blastocyst (Section 1.4.1), are usually described as pluripotent and can become any of the approximately 200 different types of cell derived from the three different germ layers. In adults, stem cells are found distributed in the various tissues of the body; they are known as multipotent stem cells (MSCs) as they retain their ability to divide and differentiate throughout life, but can only produce a specific subset of cell types.

- ■ What type of cells do haematopoietic stem cells give rise to and where are they located?

- ☐ These stem cells are located within the bone marrow and give rise to various different types of blood cells.

The remarkable ability of stem cells to develop into many cell types has obvious potential for use in regenerative medicine. Indeed, haematopoietic stem cells have been long used to repopulate the bone marrow in cancer patients, the so-called 'bone marrow transplant'. More recently, umbilical cord blood has been found to be a rich source of stem cells and is now routinely used to treat a variety of blood disorders. A wider range of cell-based therapies that rely on replacement of damaged or non-functioning cells are now on the horizon. In addition, the availability of different types of cultured cells derived from human stem cells could help in the laboratory testing of drugs and the modelling of human diseases.

4.4.2 Human embryonic stem cells

As human embryonic stem cells (hESCs) were the first stem cell type to be identified, you will begin by considering their unique properties. These cells are derived from donated human ova (eggs) fertilised *in vitro* in the laboratory. Cells from very early blastocyst stage embryos are transferred to culture dishes containing a suitable growth medium. This establishes a cell culture of hESCs that can be maintained in an undifferentiated state for many generations; the cells divide and gradually spread over the surface of the culture vessel. Periodic subculturing (the removal of cells and seeding into further fresh media) prevents the cells overgrowing and maintains a healthy culture. Once the cells have been characterised as true pluripotent stem cells, they can be preserved by freezing. The characterisation process is complex but relies on identifying various properties unique to hESCs. One useful marker of undifferentiated cells is the presence of a number of transcription factors (Book 1, Section 6.3.2), including NANOG, SOX2 and OCT4.

■ What is a transcription factor?

☐ A protein that binds to specific DNA sequences and regulates gene transcription (Book 1, Section 6.4).

NANOG, SOX2 and OCT4 are typically only expressed in pluripotent cells and they, in turn, control genes that inhibit differentiation. In addition to verifying the 'stemness' of the cells, it is useful to check if the chromosomes remain undamaged; this can be done by simple microscopic examination, but allowing a sample of the cells to actually differentiate is also a good test of integrity. Detaching cells from the culture dish and growing them in suspension will usually trigger differentiation; this allows the cells to clump together forming 'embryoid' bodies. If the cells are truly pluripotent, then they will spontaneously start to differentiate, forming a variety of different cell types. In fact, by adjusting the composition of the suspension media, for example by adding a cocktail of growth factors or signalling molecules known to be involved in specific differentiation pathways (Chapter 1), it is possible to influence the differentiation process in order to produce a desired cell type.

■ What general mechanism drives the differentiation of cells?

☐ Differential gene expression. Gene expression is largely controlled by the types of transcription factors the cell expresses (Section 1.4.2).

Therefore hESCs have huge potential to act as precursors of many different types of cells. The first therapeutic trial of hESCs came in 2012 as retinal implants in the eyes of patients with Stargardt's disease, a form of macular degeneration. Trials using hESCs are not, however, without significant controversy because the process required to generate the cells necessitates the destruction of an embryo. The ethical considerations surrounding the use of hESCs are not within the scope of this module; however, the controversy generated by their use and the regulatory regimes imposed in different countries have encouraged scientists to concentrate their efforts on looking for alternatives to hESCs. Attention has now turned to the use of multipotent stem cells (MSCs), and more recently the production of induced pluripotent stem cells (iPSCs): differentiated somatic cells that have been genetically reprogrammed to return to a stem cell-like state.

4.4.3 Induced pluripotent stem cells (iPSCs)

The ability to make stem cells to order – customised, personalised, pluripotent stem cells, without the controversial use of embryo tissue or the risk of rejection of transplanted cells – is a major goal for stem cell research. The first iPSCs were created from adult mouse (murine) fibroblast cells in 2006, followed closely by human skin fibroblast iPSCs in 2007, by Shinya Yamanaka's team at Kyoto University. The team used retrovirus vectors to insert multiple copies of specific transcription factor genes not normally expressed in adult fibroblast cells.

■ What is a retrovirus?

- Retroviruses are RNA viruses that, once inside a host cell, make a DNA copy of their genome which inserts into the genome of the infected cell.

Genetically engineered retroviruses carrying the genes for four transcription factors were used to infect cells, and the transcription factor genes were transcribed from a constitutive promoter carried on the integrated viral genome. Trial and error testing of a range of transcription factors known to be expressed in pluripotent stem cells revealed that a combination of four – OCT4, SOX2, KLF4 and c-MYC – was essential to return adult murine fibroblast cells to a stem cell-like state.

Under certain circumstances *KLF4* and *c-MYC* can act as oncogenes (Section 1.2.8); *certainly* if only *KLF4* and *c-MYC* are inserted into the adult fibroblast cells then the cells develop into tumour cells. Including the *OCT4* and *SOX2* genes ensures that iPSCs are produced instead. This suggests that the oncogenic properties of *KLF4* and *c-MYC* are balanced by *OCT4* and *SOX2* in some way. Unfortunately, although this process does indeed return the fibroblasts to a stem cell-like state, the potential to cause cancer makes the risks of using these cells for therapeutic purposes very significant. Additionally, the random integration of the retrovirus into the fibroblast genome at multiple points can give rise to mutations that may also promote oncogenesis (tumour cell formation), thus limiting the utility of the resulting iPSCs in clinical applications.

Safer processes for reprogramming cells using non-retroviral vectors are being sought. These include virus vectors that do not integrate their genome into the host cell's DNA, so the risk of genetic damage to the reprogrammed cell is significantly less; or plasmid vectors, like the ones used to transform bacterial cells to produce the novel proteins you encountered in Section 4.3. Mouse iPSCs were successfully generated in 2008 using a plasmid. Transcription and translation of the plasmid genes encoding the four essential transcription factors over a period of few weeks was sufficient to reprogram a small proportion of the treated cells to a stem cell-like state. Much further research is required in this area to fully determine how cells can be reprogrammed efficiently and safely; it seems that choice of the parental cell sources and the culture methods, as well as the method of reprogramming, are all of great significance.

4.4.4 Medical applications of stem cells

A long-term goal in the field of regenerative medicine is the generation of disease-specific **autologous stem cells** (i.e. patient-specific stem cells). This is a very fast-moving experimental field and in recent years the reprogramming of a variety of cell types has become increasingly routine. These cells are vitally useful as they allow insight into disease states, *in vitro* drug screening, and the opportunity to explore gene repair coupled to cell replacement therapy (Figure 4.20).

Using autologous iPSCs offers the promise of replacement of non-functional cells without the life-long use of immunosuppressive drugs to prevent tissue rejection – the inevitable consequence of all other types of transplant or graft. One very exciting potential use of iPSCs is for therapies related to Parkinson's disease. In Chapter 3 you were introduced to this chronic progressive neurodegenerative disorder, characterised primarily by major loss of dopaminergic neurons.

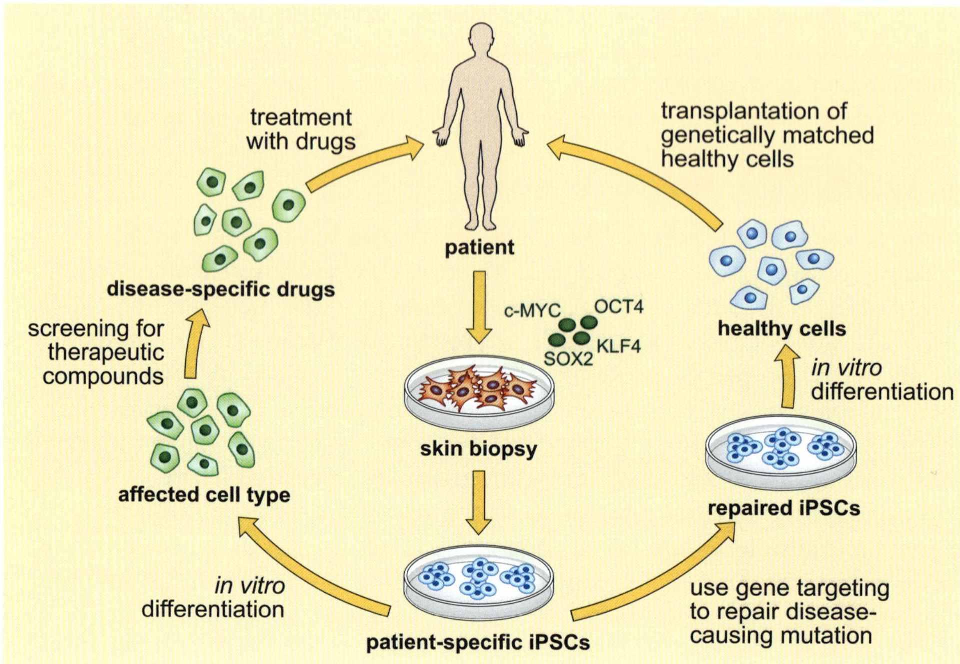

Figure 4.20 Potential medical applications of induced pluripotent stem cells (iPSCs). Reprogramming technology and iPSCs have the potential to be used to model and treat human disease. Autologous iPSCs, in this case derived by co-expression of transcription factors in cells isolated from a skin biopsy, can be used in two main ways. Patient-specific iPSCs could be directed to differentiate into the appropriate cell type for transplantation into a patient. In conditions where there is an inherited disease-causing mutation, the iPSC DNA sequence could first be repaired (right). Alternatively, autologous iPSCs differentiated into the appropriate cells type could allow the patient's disease to be modelled *in vitro* and potential drugs screened for effectiveness (left).

In Parkinson's disease, iPSC therapy offers a number of opportunities:

1   In cases of familial Parkinson's disease, where the disease-causing mutation is known, the gene could be repaired or edited in autologous iPSCs, which would then be directed to differentiate into the correct type of cell for transplantation into the patient's brain.

2   *In vitro* culture of patient-specific iPSCs directed to differentiate into dopaminergic neurons could be used to screen drugs and study the disease *in vitro*.

3   Cultured patient-specific dopamine-producing cells could also be used to replace defective dopamine-producing cells in cases of Parkinson's disease that are not linked to known genetic mutations.

It is the last of these opportunities that has captured the public's imagination. Trials in which human fetal tissue was transplanted into the brains of Parkinson's patients have shown some success in relieving symptoms of the disease. This proves the principle that dead dopamine neurons in the substantia nigra can be replaced by transplantation. Some of these grafts have given patients relief from symptoms for up to 16 years. iPSCs provide a much less controversial source of cells for this type of replacement therapy. Figure 4.21 shows some possible stem cell sources for treating Parkinson's and Table 4.2 lists some of the advantages and disadvantages of each type of cell for Parkinson's disease treatment.

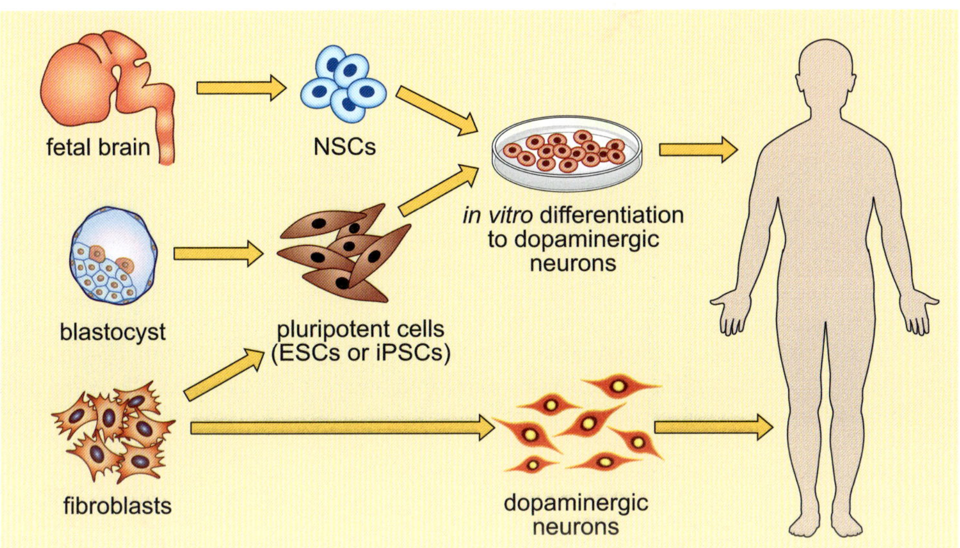

Figure 4.21 Schematic illustration of possible sources of stem cells for therapy in Parkinson's disease: (1) neural stem cells (NSCs) from human fetal brain, expanded and differentiated to dopaminergic neurons; (2) pluripotent cells generated from blastocysts (ESCs) or fibroblasts (iPSCs), expanded and differentiated to dopaminergic neurons; (3) dopaminergic neurons generated by direct conversion of fibroblasts.

There are many steps yet to overcome before stem cells can be used as a therapy for Parkinson's. It is essential that any stem cell-derived dopaminergic neurons have exactly the right characteristics. The transplanted cells must be able to release dopamine, but they must also share all the same electrophysiological and morphological properties of substantia nigra neurons. It is vital that they re-establish the network of neurons within the striatum and reconnect to other neural circuits in the substantia nigra; they must reverse the motor function deficit and most important of all, produce no adverse effects such as tumour formation or immunoreaction.

Table 4.2 Types of stem cells and their potential advantages and disadvantages in the treatment of Parkinson's disease.

Stem cell type	Definition	Advantages	Disadvantages
Embryonic stem cells (ESCs)	Pluripotent stem cells derived from the inner cell mass of the blastocyst that are able to differentiate into cells derived from all three germ layers	Highly proliferative; retain pluripotency after *in vitro* culture Shown to be able to generate dopaminergic neurons Shown to survive transplantation and generate some degree of functional recovery in humans	Risk of tumour formation Non-autologous cells so risk of rejection Ethical objections to the use of human embryos
Fetal brain neural stem cells (NSCs)	Multipotent stem cells that are able to differentiate into neurons	Lower risk of tumour formation and immune rejection than ESCs	Only limited differentiation has been achieved *in vivo* Only partial relief of Parkinson's disease symptoms Non-autologous cells, so risk of rejection Ethical objections to the use of human fetuses
Induced pluripotent stem cells (iPSCs)	Adult human fibroblasts reprogrammed into ESC-like cells	Generation of unlimited patient-specific cells for autologous transplantation Shown to survive transplantation and generate some degree of functional recovery in animals Could minimise possibility of immune reactions Remove ethical objections to treatment	Risk of tumour formation Risk of susceptibility to the original pathology of the disease

It seems that iPSCs are the most promising of these cell types so far. In 2008, dopaminergic neurons produced from mouse iPSCs were shown to alleviate symptoms when transplanted into the brains of rats with Parkinson's-like disease. In 2009, dopaminergic neurons produced from adult human fibroblasts were found to restore limited motor function in affected rats. However, before proceeding with human trials using transplantation of stem cell-derived dopaminergic neurons, the risks of tumour formation must be proven to be minimal.

After this brief introduction to the therapeutic uses of stem cells, you should now complete Activity 4.3 to further investigate their potential benefits in the treatment of neurodegenerative disease.

Activity 4.3 Parkinson's and stem cells

(LO 4.5) Allow 45 minutes to access the online resources, to read the associated notes and answer the questions in this activity

In this activity you will access a number of online resources to further develop your understanding of stem cells and the potential they may have in the treatment of degenerative diseases such as Parkinson's.

Summary of Section 4.4

- Stem cells are the precursors to all cells in the body; they can be described as either totipotent (cells of the zygote), pluripotent (e.g. human embryonic stem cells, hESCs), induced pluripotent (iPSCs), or multipotent (e.g. haematopoietic stem cells).

- hESCs can be maintained in an undifferentiated state in cell cultures indefinitely. Undifferentiated cells are characterised by the expression of specific transcription factors.

- The use of cells derived from human embryos is controversial: attention has now turned to the use of multipotent and induced pluripotent stem cells (iPSCs) for regenerative therapies.

- iPSCs are produced when somatic cells are reprogrammed to a stem cell-like state using retroviruses to insert genes for specific transcription factors. Use of retroviruses is potentially dangerous and non-retroviral reprogramming techniques are being investigated and trialled.

- The potential use of stem cells in regenerative medicine is considerable: the ability to produce autologous replacement cells for diseases such as Parkinson's and diabetes is very promising.

4.5 Final word

This module has taken you on a journey through our current understanding of cell biology, from the origins of the first cells on Earth in Book 1 to some of the latest advances in applications of cell technology in this final chapter.

Robert Hooke first described what we now know as cells in 1665. The subsequent development of increasingly powerful microscopes and ever more sophisticated techniques for labelling and analysing cells has gradually revealed their startling variety and complexity. The last few decades, since the discovery of DNA and the molecular basis of inheritance, have brought about an astonishing revolution in the understanding of cell structure and function. Modern cell biologists are able to compare the genomes of quite different organisms and reveal new links in the evolution of their fundamental cellular processes, from gene expression and reproduction, to communication and the capture and use of energy. This understanding offers the promise of new cell technologies that could improve human health, and make a prominent contribution to global economies though the industrial manufacture of biomolecules including pharmaceuticals, foods and biofuels.

We hope that you have found your studies during this module enjoyable and intriguing. You have really only scratched the surface of this fascinating topic, but the thorough understanding of cellular processes and diversity that you have gained from the module will equip you to progress to further study in cell and molecular biology.

4.6 Learning outcomes

4.1 Describe the structure and the mode of action of antibiotics and some of the resistance mechanisms to antibiotics shown by bacteria.

4.2 Explain the importance of developing alternative types of antibacterial compounds.

4.3 Describe the methods by which gene technology can be harnessed to produce useful therapeutic substances.

4.4 Outline the process technology required to mass-produce useful substances from cultured cells.

4.5 Describe the potential of stem cell technology to revolutionise the treatment of certain diseases.

References

Barber, M. (1960) 'Drug-resistant staphylococcal infection' in 'The use and abuse of antibiotics', *Journal of Obstetrics ans Gynaecology,* **67**, pp. 727–32.

Garrod, L. P. and O'Grady, F. W. (1968) *Antibiotic and Chemotherapy,* 2nd edn, Edinburgh, E & S Livingstone.

Politis, O. and Lindvall, O. (2002) 'Clinical application of stem cell therapy in Parkinson's disease', *BMC Medicine,* **10**, 1.
http://www.ncbi.nim.nih.gov/pmc/articles/PMC3261810/
Review also available at www.ncbi.nih.gov/pubmed/22216957

Acknowledgements

Grateful acknowledgement is made to the following sources:

Figures 1.7, 1.8, 1.14, 1.19: Adapted from Alberts, B. et al. (1994) *Molecular Biology of the Cell*, Taylor & Francis Inc.; Figure 1.22: Wolpert, L. and Jessell, T. et al. (1998), *Principles of Development*, Oxford University Press; Figure 1.23: Taken from www.scq.ubc.ca, used under a Creative Commons Attribution, non-commercial share-alike 2.5 Generic Licence, http:// creativecommons.org/licenses/by-nc-sa/2.5/; Figure 2.1: Dr Jiri Snaidr, Vermicon AG, Germany, www.vermicon.com; Figures 2.3, 2.9, 2.13, 2.20b: Madigan, M.M. et al. (2009) *Brock Biology of Microorganisms*, 12th edn, Pearson Education Inc.; Figures 2.4 and 3.13a: Madigan, M.M. et al. (2000) *Brock Biology of Microorganisms*, 9th edn, Pearson Education Inc.; Figure 2.5: Courtesy of Karl Stetter; Figure 2.6: Dalhus, B. et al. (2002) 'Structural basis for thermophilic protein stability: structure of thermophilic and mesophilic malate dehydrogenases', *Journal of Molecular Biology*, vol. 318. Elsevier Science Limited; Figure 2.7: Prof. Gordon T. Taylor, Stony Brook University USA; Figure 2.8: Dr David D. Wynn-Williams, British Antarctic Survey; Figure 2.12: Pikuta, E.V. and Hovver, R.V. (2007) 'Microbial extremophiles at the limits of life', *Critical Reviews in Microbiology*, vol. 33, Informa Healthcare; Figure 2.14: Mrozik, A., Piotrowska-Seget, Z. and Labuzek, S. (2004) 'Cytoplasmic bacterial membrane responses to environmental perturbations', *Polish Journal of Environmental Studies*, vol. 13, no. 5; Figure 2.15: Professor Abraham Minsky; Figure 2.18: Dr Jeremy Burgess/Science Photo Library; Figure 2.19: Udvardi, M.K. and Day, D.A. (1997) *Annual Review of Plant Physiology and Plant Molecular Biology*, vol. 4; Figure 2.20a: Canter-Lund and Lund (1995) 'Their microscopic world explored', *Freshwater Algae*, Biopress, Bristol; Figure 2.21: Woods Hole Oceanographic Institution; Figure 2.22: Tunnicliffe, V. (1992) 'Hydrothermal-vent communities of the deep sea', *American Scientist*, vol. 80, Sigma Xi, the Scientific Research Society; Figure 2.23: Visuals Unlimited, Inc./David Fleetham; Figure 2.26: Bluth, B.J., Frew, S.E. and McNally, B. (1997) 'Cell-cell communication and the lux operon in *Vibrio fischeri*', Department of Biological Sciences, Carnegie Mellon University; Figure 3.2b: © 2006 Pearson Education Inc., publishing as Benjamin Cummings; Figures 3.3a and 3.5: Biophoto Associates/Science Photo Library; Figure 3.6: CDC/ Janice Haney Carr/ Jeff Hageman, M.H.S.; Figure 3.7a: Loredo/iStockphoto.com; Figure 3.7b: Finlay Medical Society; Figure 3.9: CDC/ Rodney M. Donlan, Ph.D.; Janice Carr; Figure 3.10: Cos, P., Tote, K., Horemans, T. and Maes, L. (2010) 'Biofilms: An extra hurdle for effective antimicrobial therapy', *Current Pharmaceutical Design*, vol. 16, no. 20, July 2010; Bentham Science Publishers; Figure 3.11a: EM Unit, VLA/Science Photo Library; Figure 3.11b and 3.11f: NIBSC/Science Photo Library; Figures 3.11c left and 3.11e: Eye of Science/Science Photo Library; Figure 3.11c right: Biozentrum, University of Basel/Science Photo Library; Figure 3.11d: Kaiser, D. (2000) in Madigan, M.M. et al., *Brock Biology of Microorganisms*, 9th edn, Pearson Education Inc.; Figure 3.11g: Institut Pasteur; Figure 3.11h: Ryter, A. (1995) in Stanier, R. et al. (eds) *General Microbiology*, 5th edn,

Module team

Module team chair and author

Robert Saunders

Curriculum managers

Viki Burnage
Tracy Finnegan

Curriculum assistants

Becky Efthimiou
Dawn Partner

Book 1 chair and author

Carol Midgley

Book 2 chair and author

Jane Loughlin

Book 3 chair and author

Colin Walker

Authors

Jill Saffrey
Kerry Murphy
Diane Butler

Critical readers

Christine Gardener
Marion McKinnon

External assessor

Richard Tunwell

Production team

Greg Black (Media Developer)
Martin Chiverton (LTS Producer)
Corinne Cole (Media Assistant)
Roger Courthold (Media Developer)
Ruth Drage (Media Project Manager)
Phil Gauron (Producer, Clear Focus Productions)
Sara Hack (Media Developer)
Chris Hough (Media Developer)
Phil Howe (Media Assistant)
Richard Howes (Media Coordinator)
Martin Keeling (Media Assistant, Rights)

Bina Sharma (Media Developer)
Andy Sutton (Media Developer)
Peter Twomey (Media Developer)

Indexer

Jane Henley

Other contributors

Margaret Swithenby
Sarat Vatsavayai

Index

Entries and page numbers in **bold** refer to glossary terms. Page numbers in *italics* refer to entries in figures or tables.

X

Y

Z